ANIMAL AND PLANT

Anatomy

VOLUME CONSULTANTS

- Glen Alm, *University of Guelph, Ontario, Canada* • Erica Bower, *Botanical writer and researcher*
- John Gittleman, *University of Virginia, VA* • Tom Jenner, *Academia Británica Cuscatleca, El Salvador*
- Sally-Anne Mahoney, *Bristol University, England* • Andrew Methven, *Eastern Illinois University, IL*
- Richard Mooi, *California Academy of Sciences, San Francisco, CA* • Ray Perrins, *Bristol University, England*
- Erik Terdal, *Northeastern State University, Broken Arrow, OK*

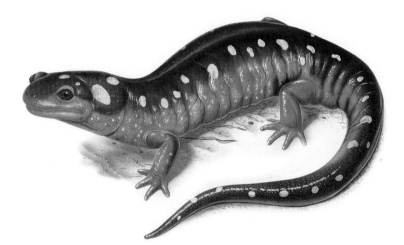

6

Mandrill – Otter

 **Marshall Cavendish**
Reference
New York

CONTRIBUTORS

Roger Avery; Richard Beatty; Amy-Jane Beer; Erica Bower; Trevor Day; Erin Dolan; Bridget Giles; Natalie Goldstein; Tim Harris; Christer Hogstrand; Rob Houston; John Jackson; Tom Jackson; James Martin; Chris Mattison; Katie Parsons; Ray Perrins; Kieran Pitts; Adrian Seymour; Steven Swaby; John Woodward.

CONSULTANTS

Barbara Abraham, Hampton University, VA; Glen Alm, University of Guelph, Ontario, Canada; Roger Avery, Bristol University, England; Amy-Jane Beer, University of London, England; Deborah Bodolus, East Stroudsburg University, PA; Allan Bornstein, Southeast Missouri State University, MO; Erica Bower, University of London, England; John Cline, University of Guelph, Ontario, Canada; Trevor Day, University of Bath, England; John Friel, Cornell University, NY; Valerius Geist, University of Calgary, Alberta, Canada; John Gittleman, University of Virginia, VA; Tom Jenner, Academia Británica Cuscatleca, El Salvador; Bill Kleindl, University of Washington, Seattle, WA; Thomas Kunz, Boston University, MA; Alan Leonard, Florida Institute of Technology, FL; Sally-Anne Mahoney, Bristol University, England; Chris Mattison; Andrew Methven, Eastern Illinois University, IL; Graham Mitchell, King's College, London, England; Richard Mooi, California Academy of Sciences, San Francisco, CA; Ray Perrins, Bristol University, England; Kieran Pitts, Bristol University, England; Adrian Seymour, Bristol University, England; David Spooner, University of Wisconsin, WI; John Stewart, Natural History Museum, London, England; Erik Terdal, Northeastern State University, Broken Arrow, OK; Phil Whitfield, King's College, University of London, England.

Marshall Cavendish
99 White Plains Road
Tarrytown, NY 10591–9001

www.marshallcavendish.us
© 2007 Marshall Cavendish Corporation

Library of Congress Cataloging-in-Publication Data
Animal and plant anatomy.
    p. cm.
    ISBN-13: 978-0-7614-7662-7 (set: alk. paper)
    ISBN-10: 0-7614-7662-8 (set: alk. paper)
    ISBN-13: 978-0-7614-7669-6 (vol. 6)
    ISBN-10: 0-7614-7669-5 (vol. 6)
    1. Anatomy. 2. Plant anatomy. I. Marshall Cavendish Corporation. II.
Title.

    QL805.A55 2006
    571.3--dc22

                                        2005053193

Printed in China
09 08 07 06 1 2 3 4 5

MARSHALL CAVENDISH
Editor: Joyce Tavolacci
Editorial Director: Paul Bernabeo
Production Manager: Mike Esposito

THE BROWN REFERENCE GROUP PLC
Project Editor: Tim Harris
Deputy Editor: Paul Thompson
Subeditors: Jolyon Goddard, Amy-Jane Beer, Susan Watts
Designers: Bob Burroughs, Stefan Morris
Picture Researchers: Susy Forbes, Laila Torsun
Indexer: Kay Ollerenshaw
Illustrators: The Art Agency, Mick Loates, Michael Woods
Managing Editor: Bridget Giles

Contents

Mandrill

ORDER: **Primates** FAMILY: **Cercopithecidae**
GENUS: *Mandrillus*

Mandrills are the largest monkeys. They live in the tropical forests of West Africa but spend most of their time on the ground. Mandrills are social animals; they are active by day and feed on a mixture of fruit, leaves, and other plant material supplemented with insects and meat from other vertebrates. Male mandrills have dramatic red and blue facial coloring.

Anatomy and taxonomy

Scientists categorize all organisms into taxonomic groups based partly on anatomical features. Mandrills are Old World monkeys belonging to the family Cercopithecidae, which includes almost 100 species living in Africa and Asia.

● **Animals** Animals are multicelled organisms that rely on organic material produced by other organisms for their nutrition. Unlike plants and some bacteria, they cannot synthesize their own food. Animals are sensitive and responsive to their environment, and most are able to move from one place to another using muscles.

● **Chordates** Animals in this large group have a bilaterally symmetrical body with a head at one end. At some point in the life history, the body is supported lengthwise by a stiff rod called the notochord.

● **Vertebrates** In vertebrates—the group that includes fish, amphibians, reptiles, birds, and mammals—the notochord is replaced early in the animals' development by a proper backbone, made of either cartilage or bone.

● **Mammals** The class Mammalia, or mammals, includes almost 5,000 known species of warm-blooded vertebrates. They are unique among animals in that females nourish their young with milk secreted from mammary glands. Mammals are hairy or furry, and their lower jawbones articulate directly with the skull.

● **Primates** The primates are an ecologically diverse but anatomically conservative group of animals, including the

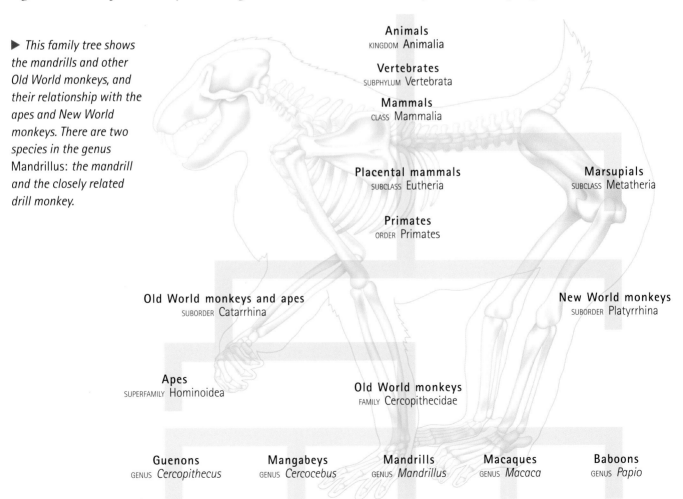

▶ *This family tree shows the mandrills and other Old World monkeys, and their relationship with the apes and New World monkeys. There are two species in the genus* Mandrillus: *the mandrill and the closely related drill monkey.*

Animals
KINGDOM Animalia

Vertebrates
SUBPHYLUM Vertebrata

Mammals
CLASS Mammalia

Placental mammals
SUBCLASS Eutheria

Marsupials
SUBCLASS Metatheria

Primates
ORDER Primates

Old World monkeys and apes
SUBORDER Catarrhina

New World monkeys
SUBORDER Platyrrhina

Apes
SUPERFAMILY Hominoidea

Old World monkeys
FAMILY Cercopithecidae

Guenons
GENUS *Cercopithecus*

Mangabeys
GENUS *Cercocebus*

Mandrills
GENUS *Mandrillus*

Macaques
GENUS *Macaca*

Baboons
GENUS *Papio*

familiar great apes and monkeys, the secretive arboreal lorises, and the lithe-bodied lemurs. Primates are divided into two basic groups: the strepsirrhines of Africa and Asia (including the lemurs, lorises, and bushbabies) and the haplorrhines, or simians, which include monkeys and apes. The haplorrhines are split into two main groups: the platyrrhines, or New World monkeys, of South and Central America; and the catarrhines, or Old World anthropoids (monkeys and apes), of Africa and Asia.

The New World monkeys include the families Cebidae and Callitrichidae, which are native to tropical South or Central America. All species have a long tail used for balancing—or, in several cases, for grasping—and they lack rump callosities (sitting calluses on the rump). New World monkeys include woolly monkeys, howler monkeys, marmosets, tamarins, and squirrel monkeys.

The Old World monkeys, or Cercopithecidae, differ from their cousins in several respects. On average they are larger, and they spend more time on the ground. One of the most significant differences is the shape of the nose, which is narrow with closely spaced, downward-facing nostrils.

● **Old World monkeys** There are almost 100 species of Old World monkeys living in Africa and Asia. Often they are treated as two separate groups: the colobines and the

▲ *Mandrills are the largest monkeys and are easily recognizable, especially in the case of the male, by their long snout, bright facial coloration, and deeply grooved nose ridges. The most dominant males have the brightest face colors.*

cercopithecines. The colobines include colobus monkeys, proboscis monkeys, and langurs. The cercopithecines include macaques, mangabeys, guenons, baboons and geladas, and mandrills. Macaques are widespread and adaptable monkeys, equally at home on the ground or in trees; all but the Barbary macaque (found in North Africa) have a long tail and live in Asia. Mangabeys are medium-size, dainty, often arboreal monkeys with partially webbed fingers; they are exclusively African and are the cercopithecines most closely related to mandrills. Guenons are medium-size monkeys with a rounded head, long limbs and tail, and variable fur color. Guenons live in much of Africa south of the Sahara Desert. Baboons and geladas are large, terrestrial monkeys with a doglike snout, prominent rump callosities, and a long tail. Baboons and geladas live only in Africa.

● **Mandrills** The genus *Mandrillus* has two species, the mandrill and the slightly smaller drill monkey; both are restricted to central West Africa. The mandrill is the largest Old World monkey.

FEATURED SYSTEMS

EXTERNAL ANATOMY Mandrills are large monkeys with coarse, olive-brown fur, a large head, and a short tail. Males in particular have brightly pigmented skin on the face and buttocks. *See pages 726–730.*

SKELETAL SYSTEM Characteristic features include a robust skull with heavy jaw, prominent browridges, and enlarged nasal bone. The backbone is adapted for a quadrupedal (four-legged) gait. *See pages 731–733.*

MUSCULAR SYSTEM Mandrills have a typical primate musculature, with males more powerfully built than females. *See pages 734–735.*

NERVOUS SYSTEM The central nervous system includes a large brain. Mandrills have excellent eyesight. *See pages 736–737.*

CIRCULATORY AND RESPIRATORY SYSTEMS Mandrills have a typical mammalian heart and circulation. *See page 738.*

DIGESTIVE AND EXCRETORY SYSTEMS Mandrills have a mixed diet, but their digestive system is that of a fruit eater, with a large stomach, long small intestine, and short large intestine. *See pages 739–741.*

REPRODUCTIVE SYSTEM Single young develop in a single-chamber uterus with a complex placenta. Infants undergo an extended period of dependency. *See pages 742–743.*

External anatomy

CONNECTIONS

COMPARE the colorful face and rump of a male mandrill with the dark, uniform coloration of another primate, such as a **CHIMPANZEE**.

COMPARE the opposable thumbs of a mandrill with the opposable digits of a reptile such as a **JACKSON'S CHAMELEON**.

Mandrills are large, robust monkeys with long, stout limbs and a short tail. The body is covered in coarse, olive-brown (and occasionally greenish) fur. The head is large with small ears and large, close-set eyes. A mandrill's eyes are protected by a prominent brow ridge that creates the impression of a permanent frown.

Mandrills show an unusually large degree of sexual dimorphism—that is, the physical differences between males and females are very marked. Males are very much larger than females and are more heavily built. Both sexes have a long, downward-sloping muzzle with five or six swollen parallel ridges on either side of the nose. The male's ridges are larger and much more brightly colored. A male mandrill is a rather fearsome looking animal, but appearances can be deceptive, and for the most part they are docile creatures—certainly less fractious and violent than their cousins, the baboons. In females and young males, the ridged areas of the snout are flushed with lilac or purple, and the nose is dark brownish-gray.

▶ **Male mandrill**
The male mandrill is large, heavily built, and distinguished by a brightly colored nose, a yellowish beard, and a colorful rump. The face and rump colors vary in intensity according to the status of the male.

The heavy **browridges** *give the impression of a permanent frown.*

An area of thicker fur forms a mane around the mandrill's **neck**.

short tail

The striking blue and red coloration of the **nose** *is repeated on the male's red penis and blue scrotum.*

The **beard** *is yellowish.*

The **belly** *is covered with pale cream fur.*

The mandrill's **fur** *is mostly grayish brown with a pale cream belly.*

20 inches (50 cm)

30 inches (76 cm)

The first **digit** *on both the hands and feet is opposable, allowing the mandrill to grip and manipulate objects with great dexterity.*

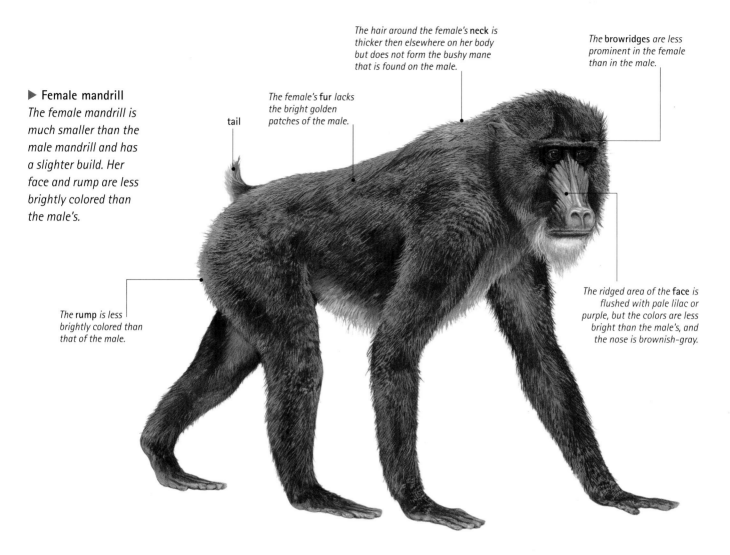

The hair around the female's **neck** is thicker then elsewhere on her body but does not form the bushy mane that is found on the male.

The **browridges** are less prominent in the female than in the male.

The female's **fur** lacks the bright golden patches of the male.

tail

▶ **Female mandrill**
The female mandrill is much smaller than the male mandrill and has a slighter build. Her face and rump are less brightly colored than the male's.

The **rump** is less brightly colored than that of the male.

The ridged area of the **face** is flushed with pale lilac or purple, but the colors are less bright than the male's, and the nose is brownish-gray.

Tails and limbs

As a general rule, all monkeys have a tail: this is one of the main features distinguishing them from apes. The tail varies between species, and that of the mandrills is unusually short—little more than a bony stub covered with coarse, tufty hair. The only truly tailless monkey, the Barbary macaque, is a distant relative of the mandrill (all other macaques have a long tail, used for balance and display). Prehensile (grasping) tails are quite rare among monkeys and apes, occurring in just four New World genera (spider monkeys, woolly monkeys, muriquis, and howler monkeys). Monkeys with a prehensile tail use it as a fifth limb to grasp branches when climbing or to pick up objects that are out of reach of the hands and feet.

Mandrills typically move on all four legs. The hind legs are longer than the forelegs; and the

EVOLUTION

Rat cousins

Most modern mammalian orders have their origins at around the same time: about 70 million years ago. Primates were then part of another group, which subsequently gave rise to several orders, including the Scandentia, or tree shrews. These shrews are small, squirrel-like mammals comprising some 19 species in a single family. They have changed little in the last 100 million years, and give us a good idea what the first common ancestor of all primates may have looked like. There is also evidence that the order Rodentia, which includes rats, squirrels, and beavers, is closely related to primates.

727

► HANDS AND FEET

The hands (upper row) and feet (lower row) of primates are broadly similar. Macaques, baboons, and guenons are closely related to mandrills and all have long digits for grasping objects such as branches and fruit. However, macaques have stronger thumbs than guenons, and baboons are more suited to standing on the ground. The hand of a colobus is slightly different and has only a short stump for a thumb. Their tree-living lifestyle and diet of leaves do not require much manual dexterity.

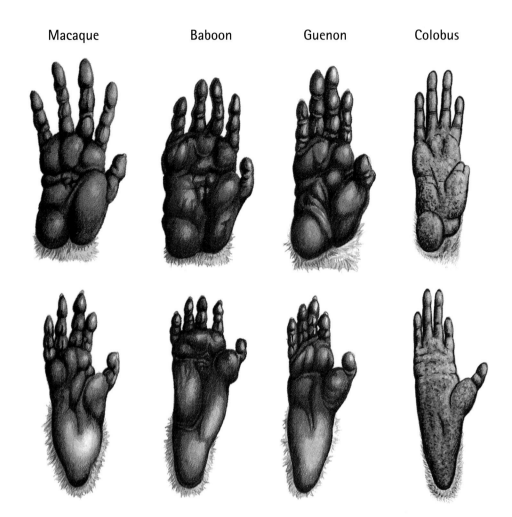

Macaque   Baboon   Guenon   Colobus

CLOSE-UP

Face paint

The face of a male mandrill is extraordinary. The ridges on either side of the nose are blue and purple, and the nose itself is bright scarlet. The effect is made all the more dramatic with the addition of pale cheek patches and a mustard-yellow beard in some animals. It is probably no coincidence that the red nose stripe flanked by blue swellings is reminiscent of the male mandrill's red penis and blue scrotum: a dominant male with well-developed facial coloring bears on his face a permanent symbol of his status and virility for all to see. Facial colors are especially bright in dominant males. Female mandrills have similar patterns and colors of facial pigmentation, but the colors are not as bright as in males.

► *The long muzzle of the male mandrill accommodates his large canine teeth. The blue bulges are outgrowths of the nose bone.*

The **dusky titi monkey** *lives primarily on leaves but will also eat fruit and insects.*

Red howler monkeys *forage for food in the middle and lower parts of the forest canopy where they eat leaves and fruit.*

▶ CEBID MONKEYS

Cebid monkeys are New World monkeys that have evolved many different lifestyles. Digestion of food such as leaves that contains a lot of cellulose requires a large stomach and therefore a large body. Therefore, cebids that rely on leaves to form most of their diet are usually larger than cebids that eat mostly insects and berries.

The **douroucouli** *eats during the hours of darkness and thus avoids many predators. Its large eyes enable it to see well in the dark.*

digits (fingers and toes) bear short, flat nails rather than claws. The first digit on each limb—the thumb, or pollex; and the big toe, or hallux—are opposable, so the mandrill can use its hands and its feet for gripping. Contrary to popular belief, opposable thumbs are not universal among primates. In New World cebid monkeys, the big toes are opposable but the thumbs are not; and colobus monkeys are unusual in that they have no proper thumb, just a vestigial stump.

COMPARATIVE ANATOMY

Varied lifestyles

With very minor physical adaptations, different primate groups have adapted to dozens of very different lifestyles, from arboreal fruit or leaf eaters to terrestrial omnivores with complex social behavior. Some primates are active by day and some by night; they climb, swing from the trees, walk on four legs or on two legs; they live alone or in groups; and they eat everything from leaves and fruit to ants and grubs— and even other primates.

The **black spider monkey** *uses its prehensile tail when foraging for food, especially among thin branches. The tail acts as an extra limb and enables the monkey to access food safely that might be difficult for other species of monkeys to reach.*

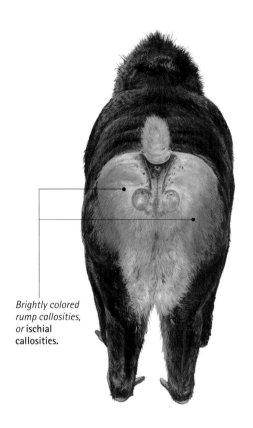

Brightly colored rump callosities, or ischial callosities.

◀ RUMP
Male mandrill
The male mandrill's rump has brightly colored hard pads called ischial callosities. As is the case with the male mandrill's face, the intensity of coloration on the rump is an indication of testosterone level. Testosterone is related to status and the higher the status of a male, the greater his level of testosterone and the brighter the colors on his rump and face.

Names and noses

Zoologists sometimes use the names "platyrrhine" and "catarrhine" to refer to New World and Old World monkeys and apes. These names refer to the structure of the nose and position of the nostrils. Platyrrhine, meaning "flat-nosed," describes the typical arrangement in New World cebid and callitrichid monkeys: the nose is very flat, with widely spaced nostrils opening out to the sides. Catarrhine, meaning "down-nosed," is the Old World arrangement: the nose is narrow, with closely spaced nostrils opening downward.

The nose of the mandrill, which protrudes slightly from the face, is the most ornate of all, flanked with ridges of bone and flamboyant skin pigmentation. Protruding noses, like those of human beings as well as mandrills, are rare among mammals, but they occur in several other species of catarrhine primates, including the snub-nosed monkeys and the proboscis monkey, which is named for its enormous pendulous nose.

Hairless skin

Like all cercopithecines, the mandrill is furry all over except for the face, palms, and soles of the feet. The buttocks are also naked, being covered in tough skin patches called callosities, which are used for sitting on. Rump callosities are seen in several other Old World monkeys, and also in apes, but they are absent in New World primates. In mandrills of both sexes, the skin of the rump callosities is brightly colored, usually a vivid shade of pink edged with lilac or purple. Zoologists think that this colorful backside helps mandrills spot one another when moving in groups through dense, gloomy forest, and a colorful rump is also indicative of high status. Mandrills are by no means the only monkeys to use color this way: for example, male vervet and patas monkeys have a bright blue scrotum, which is displayed to maximum effect when the animals stand upright on two legs during encounters with potential rivals.

The hairless skin on the hands and feet of mandrills and other primates bears ridges for generating friction. The ridges are arranged in characteristic swirls, whose precise pattern is unique to the individual. The swirls in human

▶ *The mandrill is not the only monkey to have brightly colored areas of skin. Male vervet monkeys also display bright colors. Their blue scrotum is used in displays of dominance over other males. It is also displayed to neighboring groups of vervets as a means of demarcating territory.*

fingerprints are of course used to identify individuals, but those that appear on the other padded parts of our hands or on our feet are also unique, and the same is true for all other primates.

Skeletal system

Primate skulls are typically rounded, with a short snout, a large braincase, and large eye orbits. The lack of a long snout in most species reflects the lack of highly developed smelling organs in the group as a whole. Primates that are active mainly at night have huge eyes. In nocturnal tarsiers, for example, the eyeballs are bigger than the brain, so the eye sockets are larger than the braincase.

The skull of a mandrill is distinguished by the prominent browridges and swollen nasal bones that contribute to the species' extraordinary facial appearance. The braincase is relatively large by monkey standards but smaller than that of a great ape. The eye sockets are also large but are not as exaggerated as those of nocturnal primates.

Axial skeleton

The axial skeleton consists of the skull, backbones, and ribs. The skull is connected to the rest of the skeleton by the first bones in the spine, the cervical (neck) vertebrae. The first vertebra is called the atlas bone. This is little more than a simple ring through which the spinal cord passes on its way to connect with the base of the brain. In monkeys, as with all predominantly quadrupedal mammals, the atlas bone connects with the back of the base of the skull: the skull is not balanced on top of the spine as it is humans. Following the atlas bone are six further cervical vertebrae, making up the standard seven bones seen in almost all mammals. Next are the 12 thoracic or chest vertebrae, which support the rib cage. The ribs

COMPARE the size of a mandrill's braincase with that of a great ape such as a *HUMAN*. The braincase of a mandrill is relatively large but not as large as that of a great ape.

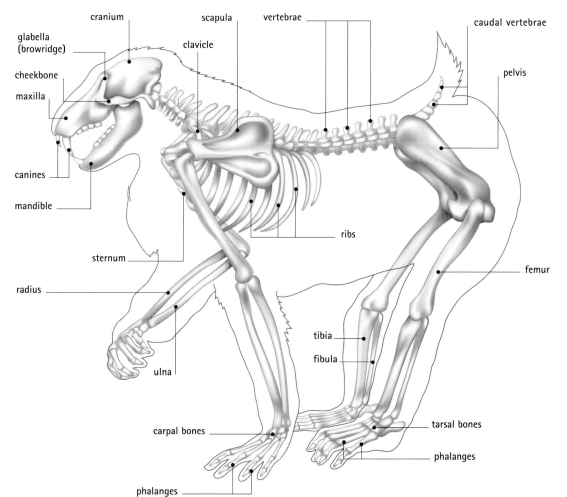

◀ **Mandrill**
The mandrill has a sturdy skeleton suited to its mostly ground-dwelling lifestyle. As is the case with other primates, mandrills are able to rotate their limbs from the elbow or knee, providing great flexibility of movement.

of monkeys are typically straighter than those of great apes, and the sternum, or breastbone, is insubstantial. Thus the chest is much narrower (deeper than it is wide) than that of a great ape and more like the chest of most other mammal groups.

All of the 12 thoracic and the 7 lumbar (back) vertebrae bear a prominent bony projection called the processa spinalis, which provides an attachment point for the major trunk muscles. Mandrills and other monkeys in the family Cercopithecidae have three pelvic vertebrae fused into a single strong unit called the sacrum, which supports the pelvic girdle. The tail in most Old World monkeys is long, with up to 17 simple caudal vertebrae; that of the mandrill, however, is much shorter.

Mandrills have a clavicle, or collarbone, and all four limbs have one upper bone (the humerus in the arms and femur in the legs)

► **MANDRILL SKULL**
The mandrill skull has a long muzzle with a set of very large canine teeth used in dominance displays. Chewing power is provided by a long row of heavy cheek teeth set in a massive jaw.

◄ **MARMOSET SKULL**
Marmosets have a relatively large brain for their size, and this is reflected in the large braincase. Marmosets are omnivorous and their dentition is suited to eating a variety of foods. Their long incisors enable them to gouge holes in trees to extract the sap.

► **COLOBUS MONKEY**
In contrast to the mandrill's skull, the skull of a colobus monkey is relatively compact, and the canines are fairly small. The incisors are used to cut vegetation, and the molars are used to mash plant matter.

and two lower limb bones: the radius and ulna in the arms and the tibia and fibula in the legs. In many other mammalian orders, the clavicle is reduced or absent, and the lower limb bones are fused to give increased strength. However, the limbs of non-primates, while very strong, are of rather limited usefulness. They can support the animal's weight, carry it (often at great speed), and perhaps be used for digging or fighting. Primates, on the other hand, can swing from trees, walk upright, manipulate objects in any orientation, reach in all directions, throw and catch, and perform acrobatics that would be quite impossible for other animals. The advantage of retaining two lower limb bones, as in primates, is that the limb is able to rotate from the elbow or knee, and this gives enormous flexibility.

Walking on hands and feet

Like baboons and many other quadrupedal monkeys, the mandrill has forelegs that are slightly shorter than the hind legs. However, when moving about on four legs, mandrills have a level posture because they support themselves on the soles of their feet and the fingers of the hands. This stance is called digitigrade for the forelimbs and plantigrade for the hind limbs. Other monkeys are fully plantigrade and walk with the palms of their hands flat on the ground. However, great apes, such as gorillas and chimpanzees, support themselves on the backs of their fingers, a form of digitigrade gait called knuckle-walking.

COMPARATIVE ANATOMY

Snouts, teeth, and noses

The long snout of mandrills and many other Old World species is reminiscent of that of a dog, but it serves a different purpose. Anatomists call it a dental snout, because it has evolved to accommodate large canines and a long row of cheek teeth, plus the associated chewing muscles. Thus the mandrill's long snout serves a different purpose from the elongated olfactory (smelling) snout of dogs or lemurs, in which the large size is associated with large nasal organs.

EVOLUTION

The origins of manual dexterity

Primate bodies have evolved from the same basic plan as all other terrestrial vertebrates. Unlike many groups, however, they have retained all five digits of the basic pentadactyl limb as separate fingers and toes. In most species these are fully prehensile, often with an opposable thumb or big toe. This is the ideal arrangement for grasping. Early primate ancestors would have used their hands for catching fast-moving invertebrate prey, much as tree shrews do today. Over time, primate hands have evolved into multipurpose tools suitable for holding young, grooming, climbing, picking and carrying food, and manipulating other objects, including tools.

▼ A mandrill feeding on a piece of fruit uses its fingers and opposable thumbs to hold and manipulate the food. This action enables the mandrill to bring the food to its mouth and bite off small chunks.

Muscular system

Muscles are specialized animal tissues capable of sustained contraction. All voluntary movements, such as those involved in locomotion and in gathering and eating food, are the result of muscle contractions. Muscles are also responsible for essential involuntary movements such as breathing, passing food through the digestive system, and pumping blood around the body.

Muscle types
The muscles of the heart have to work 24 hours a day throughout the life of the animal, so they do not tire, and they require no conscious effort to maintain activity. All other vertebrate muscles fall into one of two other types, called smooth and striated muscles. Smooth muscles are made up of short cells and are capable of contractions that are relatively limited in size. Smooth muscle is present in the walls of large blood vessels and in the digestive system. As with cardiac (heart) muscle, smooth muscle contractions are under the control of the involuntary nervous system. Striated muscles are also called skeletal muscles because they are mostly concerned with the support and movements of the skeleton. The skeletal muscles of vertebrate animals are bilaterally symmetrical: for every group of skeletal muscles that appears on one side of the body, there is a similar group on the other side, arranged in a mirror image.

Primate locomotion
The muscle systems of all primates are similar, with the basic arrangement being much the same in prosimians, monkeys, and apes. The differences between species are a matter of scale rather than structure and are related to the various forms of locomotion. For example, in brachiators (primates that move by swinging below branches) the arm muscles are

▶ Like other spider monkeys, the white-bellied spider monkey has a muscular prehensile tail that it can use as a fifth limb by curling it tightly around objects such as branches.

Antagonistic pairs

Striated or skeletal muscles can pull in only one direction. So, for movements to be reversible, muscles are arranged in what anatomists call antagonistic pairs: the contraction of one muscle moves part of the body one way, and the contraction of the other pulls it back. An example of an antagonistic muscle pair in primates is the biceps and triceps of the upper arm. The biceps lies alongside the upper arm bone, or humerus, facing forward. When it contracts, the arm flexes at the elbow. Then the contraction of another muscle—the triceps, on the other side of the humerus—is needed to straighten the arm to its original position.

◄ *When moving slowly or standing still, the mandrill distributes its weight more or less evenly on its forelimbs and hind limbs.*

stronger than those of the legs. In habitual quadrupeds, such as most monkeys, the arms and legs are equally well muscled. In humans—the only fully bipedal primates—the legs are considerably more powerful than the arms.

All fours

Mandrills are large, powerful monkeys. They habitually move around on four legs. When mandrills are walking slowly, their weight is distributed more or less evenly on the front and back limbs, although the hind limbs are rather more muscular than the forelimbs. During a gallop, the back legs are used to propel the animal in large bounds. The back legs can also be used preferentially, freeing a hand to carry an object.

Male mandrills are larger and more muscular than females. Males that are dominant in the troop are largest of all; their pumped-up physique is the result of increased levels of the steroid hormone testosterone. Once a dominant male has been deposed, his blood steroid levels drop and his muscle mass shrinks.

The facial muscles of mandrills are relatively large. Large chewing muscles move the massive lower jaw. Smaller muscles associated with facial expression are also very important. Like other monkeys, mandrills use facial expression as a means of communication.

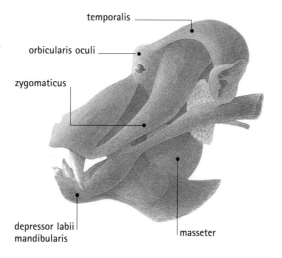

temporalis

orbicularis oculi

zygomaticus

depressor labii mandibularis

masseter

◄ **HEAD MUSCLES**
Mandrill
The muscles on a mandrill's nose cover the extensions of the nasal bone and provide the structure of nose grooves. When covered with blue skin, these grooves create a permanently snarling expression.

New World tails

Several species of monkeys, including the New World spider monkeys, have a prehensile tail, which can be used like a fifth limb for grasping objects or supporting the weight of the body when climbing. The grasping surface of a prehensile tail is typically hairless, with callused, padded skin similar to that found on other gripping surfaces such as the palms of the hands. Interestingly, there is some evidence that prehensile-tailed monkeys develop a preference for using the tail on one or the other side of the body. Just as most primates tend to be left- or right-handed, a study on spider monkeys showed that individuals could be either left- or right-tailed, but never both.

Nervous system

CONNECTIONS

COMPARE the surface folds of a mandrill's brain with the smooth surface of a **MANATEE**'s brain.

COMPARE the forward-facing eyes of a mandrill with the sideways-facing eyes of an animal such as a **RED DEER** that must be alert to the presence of predators.

The body shapes of different primate species are similar. This similarity among species is not found in other large groups of mammals. For example, the relationship between a mandrill and a human or a chimpanzee is perhaps more obvious than that between a wolf and a weasel, or a pig and a giraffe. However, if primate bodies are superficially similar, their nervous systems set them apart. The complexity and size of the brain varies greatly between primate species, as does intelligence and the relative importance of the different senses.

There is more to intelligence than brain size—after all, large animals tend to have large brains, but this does not necessarily make them smarter. New World marmosets have a very small brain, but this may be a secondary adaptation to small size rather than a sign of underdevelopment, since marmosets are probably more intelligent than many bigger-brained primates. Even relative brain size is not always an indication of intelligence: nocturnal tarsiers have relatively large forebrain relative to the size of their body, but this is mostly taken up with the unusually large visual

IN FOCUS

Seeing in stereo

Primates have forward-facing eyes with overlapping fields of vision. The image collected by one eye is similar to that collected by the other, but taken from a slightly different angle. You can see this for yourself if you stare straight ahead and wink or cover your left and right eye in turn: the image you see will appear to jump from side to side. When both eyes are open, the brain is able to blend these two sets of

information into a stereoscopic mental image with a perception of depth and distance. This ability helps primates move confidently through complex three-dimensional environments such as tree canopies, accurately judging potentially risky jumps from branch to branch. Stereoscopic vision also enables primates to catch moving objects. Mandrills can snatch a flying insect out of the air using their fingers.

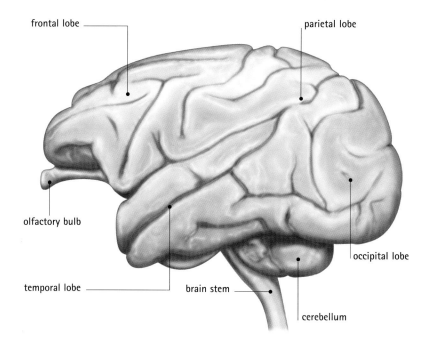

frontal lobe

parietal lobe

olfactory bulb

occipital lobe

temporal lobe

brain stem

cerebellum

Touching and sensing

The padded tips of primate fingers are especially well served with nerves and are thus highly sensitive to touch. In higher primates, unlike most other mammals, the fingers replace the snout as the main region of tactile sensitivity—with the great advantage that we have 10 of them. Primates can identify objects by touch alone, recognizing all degrees of form, texture, moisture, temperature, hardness, and malleability. As primates ourselves, we find it difficult to imagine not having sensitive fingers. Not many other animals, however, can use touch alone to recognize the subtle difference in the feel of a perfect juicy pear and one that needs an extra day to ripen.

▲ BRAIN
Mandrill
The mandrill's brain is covered with folds called gyri. Animals with extensive folding on the brain surface, or cortex, are usually more intelligent than those whose brain surface is smooth.

cortex. By most other reckonings, tarsiers are not particularly intelligent.

The cerebrum, which comprises the cerebral hemispheres, is the part of the brain that has evolved most recently. This region is associated with complex thought processes. The larger the cerebrum, the greater the degree of folding and the more complex is the characteristic pattern of grooves and folds in the surface layer of the brain (the cerebral cortex). Mammals with a relatively small, simple brain and limited intelligence have a smooth cerebral cortex. For example, the brain of prosimian tarsiers is almost completely smooth, whereas the cerebral cortex of an Old World macaque is moderately folded, but less so than that of a great ape.

Senses
For all primates, the most important sense is vision. Some scientists believe that it was the need for hand-eye coordination that first led to the increase in brain size in primates. Brain size has become a hallmark of primate evolution, leading ultimately to the outsize brain and unrivaled intelligence of great apes and especially humans. Monkeys such as mandrills are smart, but they lack the powers of reasoned intelligence that set apes apart. Monkeys' brains are smaller than those of apes, and the regions specializing in vision and control of movement are relatively large. Complex social interaction

depends on intelligence, and the brains of social cercopithecines such as mandrills and baboons are large compared with those of more solitary species. Even without a detailed examination of the eye, the garish pigments of the mandrill's face and buttocks leave little doubt that they see in color. Indeed, color vision is the norm among diurnal (day-active) primates.

Hearing in primates is adequate rather than acute but covers an extensive range: most primates can pick up high- and low-pitched sounds, an ability that gives them a very good awareness of their surroundings.

Mandrills have a scent gland on their chest that they use to mark their territory, so their sense of smell is important, though less important than either vision or hearing. Most monkeys have a reduced rhinarium. A rhinarium is the moist area of supersensitive skin around the nose of scent-oriented animals such as carnivores and ungulates.

Primates also have a typically short snout, which leaves little room for large smelling organs. Prosimians and New World monkeys have a vomeronasal organ located in the roof of the mouth, which helps them detect certain chemicals in the air. This organ is greatly reduced in Old World monkeys and apes.

Circulatory and respiratory systems

The primate heart has four muscular chambers through which blood is pumped in turn as part of its continual cycling around the body. Oxygen-depleted blood containing high concentrations of carbon dioxide (the waste product of respiration) arrives in the right atrium and is passed into the muscular right ventricle; from there it is forced under gentle pressure the short distance to the lungs. In the lungs, the blood is relieved of excess carbon dioxide and takes on freshly inhaled oxygen, which binds to molecules of hemoglobin in the red blood cells. Oxygenated blood drains from the lungs back to the heart; this time it enters the left atrium and passes into the left ventricle. The left side of the mammalian heart is more powerful than the right, and it forces blood out and around the body via a very large blood vessel, the aorta.

As a general rule, as body size increases average blood pressure increases and heart rate decreases. Primate size varies greatly, so blood pressure and heart rate are both highly variable among species. Typical systolic (maximum) pressure for mandrills and other large monkeys is around 90 mm Hg (mercury)—similar to that of a young human—and the average heart rate is around 100 beats per minute. Blood pressure usually increases as an animal ages, since large arteries lose their elasticity.

Showing their colors

The colors on the face and buttocks of a mature mandrill change to reflect the animal's condition and emotional state. Under certain circumstances the colors seem much brighter, sometimes almost glowing in the gloom of the forest shade. The pink color is the result of blood vessels passing close to the skin, and the intensity of color can vary depending on how much blood is pumped that way—just as a fair-skinned human can blush pink or red. Red blood also enhances the purple hue of the male mandrill's snout and rump patches.

COMPARATIVE ANATOMY

Hanging on tightly

Most people are familiar with the unpleasant numb feeling in the hands that results from carrying heavy shopping bags—or swinging from a trapeze. This is a result of reduced blood flow caused by pressure on the fingers. Some primates, most notably the lorises, have unusual circulation in their hands and feet, with an extensive network of additional capillaries. These fine blood vessels keep the hands and feet very well supplied with oxygenated blood so that they do not tire during prolonged grasping. As a result, lorises can hang on tightly to branches for hours on end.

▼ HEART
Mandrill
The arrows indicate the direction of flow of blood through the chambers of the heart.

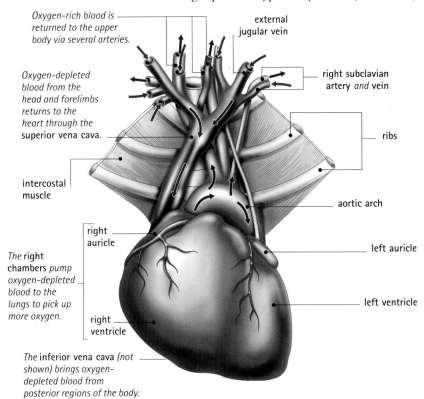

Oxygen-rich blood is returned to the upper body via several arteries.

Oxygen-depleted blood from the head and forelimbs returns to the heart through the superior vena cava.

intercostal muscle

The right chambers pump oxygen-depleted blood to the lungs to pick up more oxygen.

right auricle

right ventricle

The inferior vena cava (not shown) brings oxygen-depleted blood from posterior regions of the body.

external jugular vein

right subclavian artery *and* vein

ribs

aortic arch

left auricle

left ventricle

Digestive and excretory systems

As a group, the cercopithecines have the most generalist diets of any primates—many species eat almost equal quantities of fruit, leaves, and animal material. As a rule, only the smaller species eat a large proportion of insects—for an animal as large as a mandrill these are mere morsels and not really worth foraging for specifically. Most of the meat in the diet of large monkeys like mandrills and baboons comes from larger animals, up to the size of small antelopes such as duikers. However, hunting takes a lot of energy and thus is usually opportunistic. For the most part, mandrills eat nutrient-rich plant material such as fruit, nuts, roots, and seeds, bulked up with a variable quantity of leaves.

Teeth for all occasions

All adult cercopithecines have the same arrangement of teeth. At the front of each jaw there are two pairs of incisors flanked by a single pair of large canines. In the mandrill, the upper canines are long, pointed tusks, and the lower pair are shorter and curve backward. There is a small gap between the upper incisors and upper canines into which the lower canines settle when the mouth is closed. The cheek teeth (molars and premolars) all have cusped and ridged grinding surfaces, enabling mandrills to chew tough plant material.

Storage solutions

Mandrills, drills, and several other leaf-eating monkeys (baboons, macaques, mangabeys, and guenons) have large cheek pouches outside the teeth in which leaves or other foodstuffs are stored temporarily so that they can be carried away and eaten at leisure. Other Old World monkeys, in particular the colobines, have evolved even more specialized adaptations for leaf-eating. In colobus, langurs, and proboscis monkeys there are no cheek pouches, but the stomach is very large, with a folded, pouched internal structure in which leaves are retained for a long time, thus maximizing the efficiency of digestion.

CONNECTIONS

COMPARE the short large intestine of a mandrill with the long large intestine of a RHINOCEROS.

COMPARE the dentition of a mandrill with that of a carnivore such as a LION.

▼ INTESTINES
The small intestine is long, but the large intestine is short. This arrangement is similar to that of most fruit-eating mammals.

IN FOCUS

Leaf-eating mammals

Leaves are a very reliable source of nutrient. In the tropics there are leaves on the trees all year around, whereas fruits, seeds, and flowers tend to be more seasonal. However, while finding leaves is usually very easy, digesting them can be much more difficult. Leaves contain large quantities of cellulose, a complex form of carbohydrate that cannot be broken down by normal mammalian digestion. Mammals that specialize in eating leaves enlist the help of special cellulose-digesting bacteria, which live in the gut and break down the tough plant material into smaller constituent molecules that the animal can digest. Even with the help of these bacteria, digesting cellulose takes a long time, and so leaf eaters have a large gut. Leaf-eating monkeys are no exception.

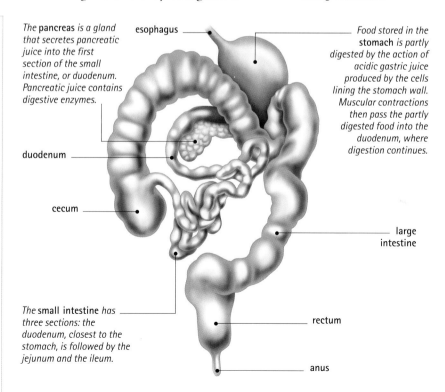

The pancreas is a gland that secretes pancreatic juice into the first section of the small intestine, or duodenum. Pancreatic juice contains digestive enzymes.

esophagus

duodenum

cecum

The small intestine has three sections: the duodenum, closest to the stomach, is followed by the jejunum and the ileum.

Food stored in the stomach is partly digested by the action of acidic gastric juice produced by the cells lining the stomach wall. Muscular contractions then pass the partly digested food into the duodenum, where digestion continues.

large intestine

rectum

anus

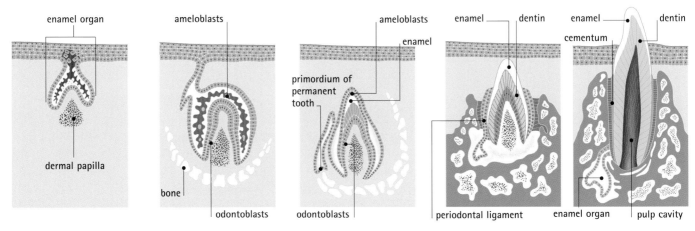

enamel organ   ameloblasts   ameloblasts   enamel   dentin   enamel   dentin

enamel

cementum

primordium of permanent tooth

dermal papilla

bone

odontoblasts   odontoblasts

periodontal ligament   enamel organ   pulp cavity

▲ TOOTH DEVELOPMENT IN A MAMMAL

In the first stage an enamel organ develops from the mouth's surface layer of skin, or epidermis, and dermal papilla appear. Second, the enamel organ produces cells called ameloblasts, which make tooth enamel, and the dermal papilla produces odontoblasts, which are the source of bonelike dentin. Third, the epidermis and dermal papilla produce a primordium. This lies next to the developing deciduous, or milk, tooth and will ultimately become a permanent tooth. The deciduous tooth continues to grow in the fourth stage, until in the fifth stage the deciduous tooth erupts through the gum. Before the deciduous tooth is lost, the primordium will begin to develop into the permanent tooth.

◀ *A mandrill displays its large canine teeth. Rather than for cutting food, mandrills use these teeth for display and for male–on–male fighting.*

the larger molecules used to build and maintain an animal's body. However, these tiny building molecules are not the only products of digestion. Every time the chemical bonds that hold a molecule together are broken, energy is released. All organisms are able to harness this energy to power the many processes that keep them alive.

In the mandrill, as in most mammals, the process of digestion begins in the mouth, where food is chewed and mixed with saliva.

The mandrill gut is basically that of a predominant frugivore (fruit-eating animal). There is a long small intestine that maximizes the uptake of easily digested sugars, and a short large intestine that allows only limited time for the digestion and absorption of complex carbohydrates such as cellulose and starch.

Digestion

In simple terms, digestion is the breakdown of organic material into smaller and smaller residues. Some of these residues, for example peptides and fats, can then be reassembled into

EVOLUTION

Color vision and frugivory

Plants have many adaptations that enable them to avoid being eaten by herbivorous animals—spines, thorns, and foul-tasting or toxic leaves are all examples of protective measures. However, animals that eat fruit rather than leaves can be beneficial to plants—by swallowing the fruit and depositing the indigestible seeds in their feces elsewhere, they help the plant disperse. Thus, over time plants have evolved fruits that are ever more appealing to fruit eaters. Being sweet and nutritious, fruits are worth looking for; and being strongly scented and often brightly colored, they are easy to find—provided that the animal concerned has a good sense of smell or color vision. It seems likely that the evolution of large colorful fruits on certain trees has occurred in tandem with the evolution of ever more discerning vision in certain primates. Larger colorful fruits enable the trees to attract more primates, and improved vision enables the primates to find the colorful fruits. As to which came first—color vision or colorful fruits—biologists can really only speculate.

Saliva not only makes food easier to swallow, but it also contains enzymes that kick-start digestion. The function of chewing is twofold: it begins the physical process of breaking up food, and it also helps spread the digestive enzymes evenly through each mouthful before food is swallowed. Once in the stomach, food is subjected to more potent chemical attack, by acidic digestive juices containing more enzymes. As food passes from the stomach to the small intestine, it is mixed with bile from the liver and pancreatic juices. Both of these secretions are highly alkaline, so they help neutralize the acidic material emerging from the stomach. Bile acts as an emulsifier, dispersing fats so they can be more easily digested. The pancreatic juices contain a range of enzymes that interact specifically with various fats, proteins, and carbohydrates, breaking them up into smaller molecules. The digestive processes that take place in the alimentary canal itself are relatively crude, and only the weaker bonds holding food molecules together are broken. Very little energy is released during this stage—and that is just as well, since energy released in the gut cannot be used to fuel the body but is dissipated as heat. The products of digestion in the gut are molecules that are small enough to be absorbed into the cells lining the gut.

Fat and muscle

Once food molecules are taken into a cell, they can be processed on the spot or exported in the bloodstream to other parts of the body. The energy released by further digestion inside cells can be used to fuel the cell or converted into a

COMPARATIVE ANATOMY

Lower jaw

Dietary specializations are often reflected in the structure of an animal's lower jaw. That of the mandrill is large and robust, supporting large cheek teeth capable of grinding up tough vegetable matter and cracking nuts. The large canine teeth are used mainly in display, but with such a powerful bite, they also make effective weapons if necessary. The colobus has highly cusped cheek teeth and a powerful jaw with large muscle attachments—ideal for shredding and grinding leaves. The langur is more of a fruit-eating specialist. Its jaws are robust, but because of the smooth shape of the skull, the associated chewing muscles are not particularly large. The marmoset jaw is more slender, and there are no millstonelike cheek teeth—these dainty monkeys do not do much heavy chewing. The clue to their specialization is the large, chisel-shaped incisors. These are used to gouge small holes in the bark of gum-producing trees so that small amounts of nutritious sap can be lapped up.

form in which it can be easily stored. Short-term energy is stored as a chemical called ATP, whose molecular bonds yield large amounts of energy when broken. If the animal is well-nourished, it can also lay down long-term energy reserves of fat. The significant increase in body size that takes place in dominant male mandrills is a result of increased muscle mass as well as fat accumulation. Increasing muscle mass is a less effective means of storing energy: a fat male mandrill may cope with starvation better than a well-muscled one, although it will not achieve the social dominance needed for successful breeding. The allocation of excess food resources to fat stores or muscle is controlled by muscles.

▼ MAMMALIAN ALIMENTARY CANAL
The alimentary canal of the mandrill digestive system has the same tissues as that of most other mammals. The canal is divided into several distinct regions, each with different functions. The small intestine, for example, has a surface covered with many tiny projects called villi that increase the surface area over which digestion can take place.

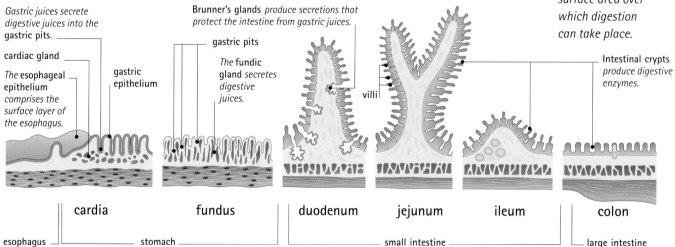

Gastric juices secrete digestive juices into the gastric pits.

cardiac gland

The esophageal epithelium comprises the surface layer of the esophagus.

gastric epithelium

Brunner's glands *produce secretions that protect the intestine from gastric juices.*

gastric pits

The fundic gland secretes digestive juices.

villi

Intestinal crypts *produce digestive enzymes.*

cardia | fundus | duodenum | jejunum | ileum | colon

esophagus | stomach | small intestine | large intestine

Reproductive system

The reproductive biology of primates is highly conservative—that is, it differs very little from species to species. Typically, very few young are born in each litter: single young are usual for most species. Females have just two teats, located high on the chest in the pectoral region, so suckling young are habitually cradled in the arms. Young primates are born in a highly dependent state. Their big brain gives them a very large capacity to learn, but learning takes time, so there tends to be an extended period of dependence.

Females have two ovaries, connected to a single uterus by the fallopian tubes. In prosimians, such as lemurs and lorises, the uterus has two distinct lobes. This arrangement is called a bicornuate uterus and represents an evolutionary halfway point between the primitive double uterus of marsupials and many rodents and bats, and the single-chamber uterus seen in higher primates.

In male mandrills, the testes descend from the body and lie in a pouchlike scrotum. One explanation for this arrangement is that it helps keep the testes a few degrees cooler than the core body temperature and thus provides optimum conditions for the production and storage of healthy sperm. (However, some biologists believe that the evidence for this explanation is weak.)

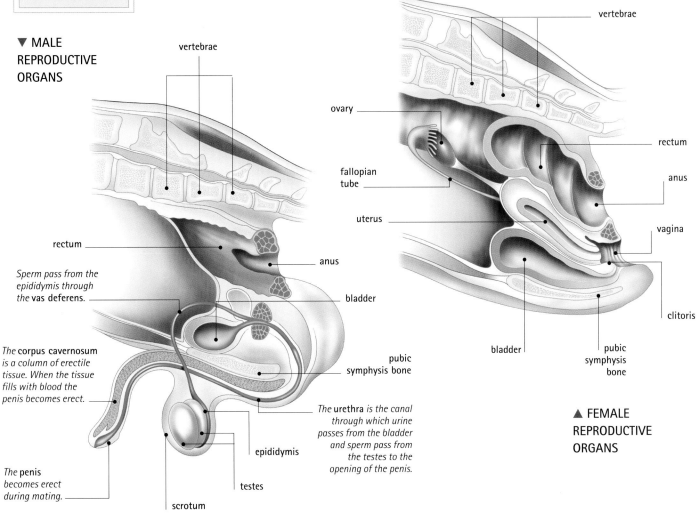

▼ MALE REPRODUCTIVE ORGANS

vertebrae

rectum

Sperm pass from the epididymis through the vas deferens.

The corpus cavernosum *is a column of erectile tissue. When the tissue fills with blood the penis becomes erect.*

The penis *becomes erect during mating.*

anus

bladder

pubic symphysis bone

The urethra *is the canal through which urine passes from the bladder and sperm pass from the testes to the opening of the penis.*

epididymis

testes

scrotum

vertebrae

ovary

fallopian tube

uterus

rectum

anus

vagina

clitoris

bladder

pubic symphysis bone

▲ FEMALE REPRODUCTIVE ORGANS

◀ A mandrill mother feeds her young. As is the case with other primates, the mammary glands are located high on the mother's chest.

Testosterone and the winning streak

Testosterone levels vary tenfold between male mandrills, and up to fivefold within a single individual over a short period of time. Males that live alone have reduced testosterone levels compared with those that live as part of a group. All males show increased testosterone in the presence of estrous females or during times of aggression or stress. Elevated levels persist for some time after a successful fight. Thus a male that keeps fighting and winning begins to exhibit significant testosterone-related characteristics such as thicker mane and beard, enhanced facial pigmentation, and a general fattening, resulting in a tough, puffed-up appearance. These changes are only semipermanent: following a series of defeats, testosterone levels decline and the animal's appearance and demeanor can change dramatically over a period of weeks.

For sperm to be delivered to the female during mating, they must pass back into the body through a series of convoluted ducts and eventually out through the urethra—the same tube that carries urine from the bladder to the tip of the penis. In a large primate, such as a human, this involves a round trip of 30 inches (80 cm) or more.

Mandrills are long-lived, often surviving for 30 to 40 years. Like most Old World monkeys, they mature slowly: females first breed at age three or four years, though some may first breed at age two. Breeding is seasonal, with females entering estrus (the period during which they are capable of conceiving) every 33 days or so until they conceive. Only dominant troop males gain access to estrous females. Gestation lasts almost six months, and young are born singly and cared for exclusively by the females.

AMY-JANE BEER

The hemochorial placenta

Mandrills and other Old World monkeys share with apes a highly specialized form of placenta. The placenta is the remarkable temporary organ that allows gases and nutrients from the maternal bloodstream to pass into the blood of the fetus. The fetus is connected to the placenta by blood vessels running through an umbilical cord. In the hemochorial placenta, the fetal blood vessels are effectively bathed in the maternal blood as they pass though the placenta, and the exchange of respiratory gases, nutrients, and metabolic waste is very effective. This specialized arrangement requires elaborate changes in the area of the uterine wall where the placenta will develop. These changes begin to take place as part of the estrous cycle even before the female is pregnant. If conception does not take place, the specially prepared uterine lining disintegrates and is shed: this is the basis of menstruation.

FURTHER READING AND RESEARCH

Ankel-Simons, F. 2000. *Primate Anatomy* (2nd ed.). Academic Press: Burlington, MA.

Dunbar, Robin, and Louise Barrett. 2000. *Cousins: Our Primate Relatives*. BBC Worldwide: London.

Nowak, R. M. 1999. *Walker's Mammals of the World* (6th ed.). Johns Hopkins University Press: Baltimore, MD.

Primate Information Network: http://pin.primate.wisc.edu/

Marsh grass

CLASS: Angiospermae   FAMILY: Gramineae
GENUS AND SPECIES: *Spartina alterniflora*

Marsh grass lives along the coastline of America. It has many of the features common to all grasses, but some special features of its anatomy help it survive in its unusual habitat of salty, flooded mudflats and estuaries. Most other grasses prefer dry habitats—but grasses are experts in survival and now live in most habitats on Earth.

Anatomy and taxonomy

All life-forms are classified in groups of relatively closely related species. The classification is based mainly on shared anatomical features, which usually indicate that the members of a group have the same ancestry. Thus the classification shows how life-forms are related to each other.

● **Plants** All true plants are multicellular (in contrast to most of the plantlike protists, which are single-celled). Unlike animals and fungi, most plants can make their own food using the energy from sunlight to turn simple chemicals (carbon dioxide and water) into carbohydrates. This process is called photosynthesis (from the Greek words *photos*, meaning light; and *syntithenai*, to manufacture). Carbohydrates are the building blocks for all the materials that make up a plant. Chlorophyll, the compound that gives plants their green color, makes photosynthesis possible by capturing the sun's energy.

● **Angiosperms** There are two major groups of seed plants: gymnosperms and angiosperms. Gymnosperms have naked seeds and include pine trees and other conifers. Angiosperms, which have enclosed seeds, are the flowering plants; they have their ovules inside an ovary, which ripens into a fruit containing the seeds.

● **Monocotyledons** Flowering plants are divided into the monocotyledons, or monocots, and dicotyledons, or dicots. Grasses are monocots; the term means that the seedling has a single first leaf or cotyledon. In contrast, dicots (such as apple trees, daisies, cotton, and the Venus flytrap) have a pair of cotyledons. Palms, lilies, orchids, and grasses are all monocots and usually have long, narrow leaves with parallel veins, rather than the netlike vein patterning of dicots.

● **Grasses (Gramineae)** Grasses tend to be overlooked, as they are generally fairly small plants and do not have showy flowers. They are, however, one of the most important plant families in the world. Cereal crops are mostly grasses, cultivated for their edible seeds. Seeds of wheat, oats, barley, rye, maize, rice, and millet form the staple food of most of the human population. Sugarcane (*Saccharum officinarum*) stems are the source of about 60 percent of the world's sugar (the other 40 percent comes from sugar beet, a dicot). We also depend on grasses as fodder for domestic animals. We even use some of the larger grasses, such as bamboos, as construction materials. There are approximately 9,500 grass species throughout the world, with the greatest diversity in

▼ This classification tree shows how botanists arrange species of flowering plants into groups based on their similarities. Botanists do not always agree on which characteristics should be used to judge similarity, so there are many different systems.

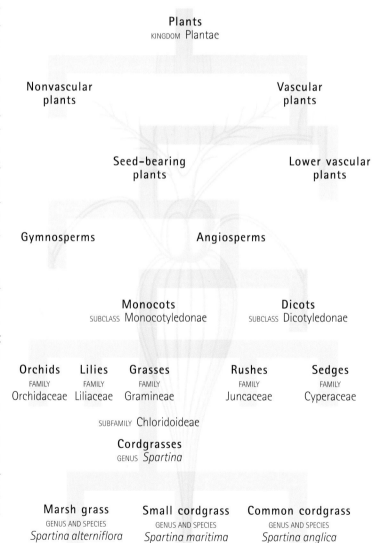

Plants
KINGDOM Plantae

Nonvascular plants

Vascular plants

Seed-bearing plants

Lower vascular plants

Gymnosperms

Angiosperms

Monocots
SUBCLASS Monocotyledonae

Dicots
SUBCLASS Dicotyledonae

Orchids
FAMILY
Orchidaceae

Lilies
FAMILY
Liliaceae

Grasses
FAMILY
Gramineae

Rushes
FAMILY
Juncaceae

Sedges
FAMILY
Cyperaceae

SUBFAMILY Chloridoideae

Cordgrasses
GENUS *Spartina*

Marsh grass
GENUS AND SPECIES
Spartina alterniflora

Small cordgrass
GENUS AND SPECIES
Spartina maritima

Common cordgrass
GENUS AND SPECIES
Spartina anglica

tropical and northern temperate regions. Grasses have a wider distribution than any other plant family. There are grasses in almost every land habitat on Earth, from high mountain steppes to deserts and icy tundra. Only lichens and algae can live in more extreme situations.

Nearly all species live in open situations—on open plains, savannas, steppes, and even marshes and tidal mudflats. Bamboos are one of the few grasses that can grow in dense forests. Most grasses are herbaceous plants, but bamboos have woody stems. The giant bamboo can grow to over 100 feet (40 m) tall. The herbaceous grasses may be tufted, tussocky, or spreading. Lawn grasses are spreading, forming an even mat when mown. Grasses that spread do so by stems that extend aboveground and

EXTERNAL ANATOMY Grasses have rounded or flattened stems that are usually hollow, with joints called nodes. The long, thin leaves are tough and grow along the length of the stem in two alternating ranks. Roots anchor the plant, and underground stems called rhizomes help it to spread. *See pages 746–748.*

INTERNAL ANATOMY In the leaves, the vascular bundles are in parallel rows. Bundle sheath cells surround them, and are where most photosynthesis takes place. *See pages 749–750.*

REPRODUCTION Marsh grass, like all grasses, is wind-pollinated. Flowers are tiny, and grouped in a thin spike. Marsh grass also reproduces vegetatively (without flowers). *See pages 751–753.*

▲ *Marsh grass grows on mudflats and estuaries and may colonize extensive areas in dense mats. Marsh grass is able to spread rapidly by vegetative (asexual) reproduction.*

belowground. Grass stems are hollow, and interrupted at intervals by swollen joints or nodes. The most familiar part of a grass is the leaves, which are always long and thin, with parallel veins.

- **Chloridoideae** There are five main subfamilies of grasses. The Chloridoideae are distinguished from the other groups of grasses by their flower structure, their use of a particular photosynthetic pathway (C4), and the associated specialized leaf anatomy. Many of the Chloridoideae are tolerant of drought and salt, and most members of this subfamily thrive in hot, dry climates. Marsh grass is an exception, as it lives in wet conditions.

- **Cordgrasses** The scientific name for cordgrasses, *Spartina*, is from the Greek *sparton*, a type of rope made of grass. Cordgrasses are perennial (surviving and growing over several years) with creeping rhizomes (underground stems), erect stems, and thick leaves. There are 17 species, living in coastal salt marshes in America, Europe, and Africa.

- **Marsh grass** Marsh grass (also called smooth cordgrass), *Spartina alterniflora*, lives in tidal areas, salt marshes, and estuaries along the U.S. Atlantic and Pacific coasts. Because there are not many other plants that can tolerate salty water, marsh grass often grows in vast swathes uninterrupted by other plant species.

External anatomy

CONNECTIONS

COMPARE the structure of marsh grass with an *APPLE TREE*. None of the grasses have the strong central trunk of an apple tree. Grasses have a jointed stem instead of a trunk.

Grasses, like most other plants, have above-ground stems and leaves, which are green and photosynthesize. At the base of the stem are the roots, which anchor the plant in the soil and conduct water and minerals from the soil to the stems and leaves.

Stems

Marsh grass stems are similar to most grass stems in that they are cylinder-shaped and hollow, with joints called nodes at intervals. Marsh grass is a tall grass, with flowering stems that can grow to 8 feet (2.4 m) in height. The stems are generally unbranched, except at the base, and up to 0.5 inch (1 cm) in diameter. They form large, erect clumps and are robust, able to support the plant's leaves and flowers even while being battered by waves.

Leaves

The leaf is the food factory of the plant. Leaves capture light energy and use it to build up carbohydrates from the atoms in water and carbon dioxide in the air. Because of the need for both light and air, leaves are usually thin, with a large surface area.

IN FOCUS

A damage-resistant design

Grasses have many features that make them well-adapted to grazing. Stems called tillers spread out away from the parent plant, often staying close to the ground, away from animals' teeth. Buried rhizomes are even more protected. Grasses have lots of growing points, and many grasses also sprout new growing points (meristems) along ordinary branches. If a few growing points are nibbled off, the plant can still grow from others. Most grasses are also tough. Their leaves contain silica (the mineral in glass), which makes them hard to chew. By surviving grazing better than their competitors, grasses can dominate in areas where grazing occurs, producing vast grasslands such as the Great Plains.

Each grass leaf consists of three parts: the sheath, the ligule, and the blade. The sheath forms the base of the leaf. It grows from a node and forms a tube enclosing the stem. There is a transition zone where the sheath turns into the blade. The ligule is an extension of the sheath at this point, while the leaf blade bends away from the stem. In marsh grass, the ligule is a fringe of hairs. In other types of grass, the ligule may be a long flap, collar-shaped, or almost absent.

As in other grasses, the leaf blade of marsh grass is long and thin—up to 18 inches (46 cm) long and 0.6 inch (1.5 cm) wide—with parallel veins. The leaves are flat (not furrowed or channeled as in some other species), and taper to a long involute tip (edges rolled, with the upper surface to the inside). The blades are smooth, hence the name smooth cordgrass. Other cordgrasses are rough to the touch owing to the presence of tiny spines or hairs.

Rhizomes

Rhizomes are underground stems. They grow through the soil, then sprout new shoots from

▼ ROOTS
Marsh grass
The roots of marsh grass are covered with tiny root hairs that greatly increase the surface area across which water can be absorbed.

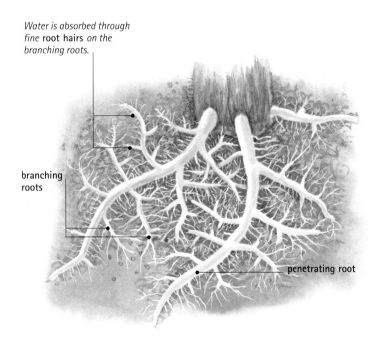

Water is absorbed through fine **root hairs** on the branching roots.

branching roots

penetrating root

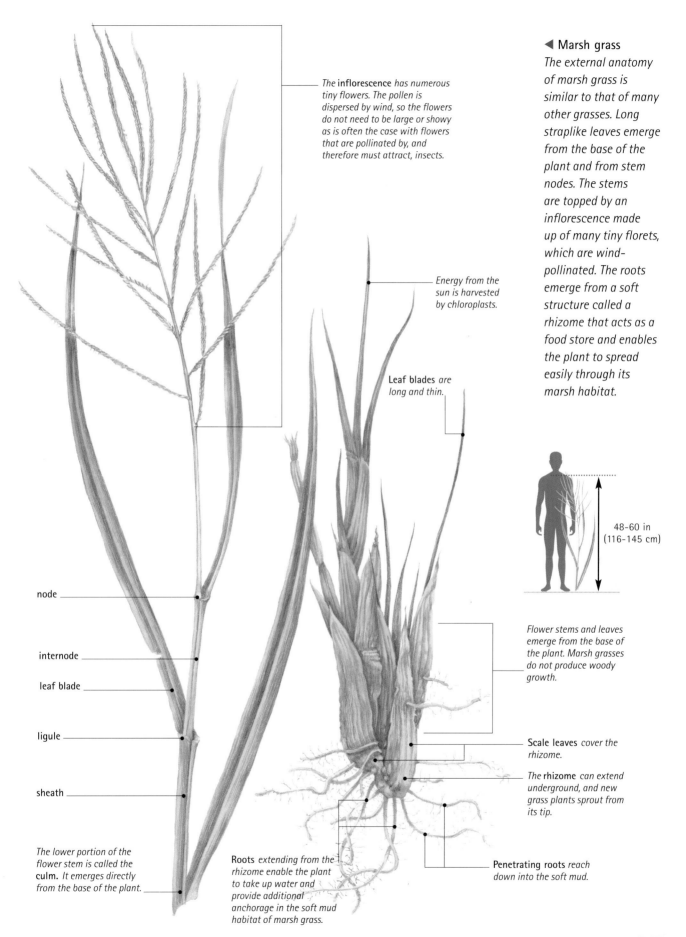

The **inflorescence** has numerous tiny flowers. The pollen is dispersed by wind, so the flowers do not need to be large or showy as is often the case with flowers that are pollinated by, and therefore must attract, insects.

◄ **Marsh grass**
The external anatomy of marsh grass is similar to that of many other grasses. Long straplike leaves emerge from the base of the plant and from stem nodes. The stems are topped by an inflorescence made up of many tiny florets, which are wind-pollinated. The roots emerge from a soft structure called a rhizome that acts as a food store and enables the plant to spread easily through its marsh habitat.

Energy from the sun is harvested by chloroplasts.

Leaf blades are long and thin.

48–60 in
(116–145 cm)

node

internode

leaf blade

ligule

sheath

Flower stems and leaves emerge from the base of the plant. Marsh grasses do not produce woody growth.

Scale leaves cover the rhizome.

The **rhizome** can extend underground, and new grass plants sprout from its tip.

The lower portion of the flower stem is called the **culm**. It emerges directly from the base of the plant.

Roots extending from the rhizome enable the plant to take up water and provide additional anchorage in the soft mud habitat of marsh grass.

Penetrating roots reach down into the soft mud.

747

Salt gland

The sodium and chloride ions from sea salt (sodium chloride) are extremely toxic to most plants, affecting osmoregulation (the water balance of cells) and the cells' ion balance. Plants living in salty environments (such plants are called halophytes) cope in a variety of ways. Marsh grass removes most of the salt it absorbs from the soil and seawater by excreting it through specialized organs. These organs, called hydathodes, are microhairs in the leaf that act as salt glands.

Plants that are capable of excreting salt have been used to improve soil. The grass *Leptochloa fusca* has been used in Pakistan and India to remove salt from soil so that the land can be used for farming. The leaves of the grass, encrusted with salt crystals, have to be disposed of elsewhere.

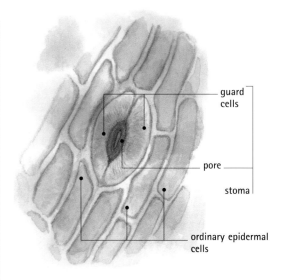

guard cells

pore

stoma

ordinary epidermal cells

▲ STOMA
Plants cells obtain a supply of carbon dioxide (CO_2) through openings called stomata (singular, stoma). The plant can also lose water through these holes. The pores of the stomata can therefore be opened and closed by guard cells on either side. This enables the plant to control water loss.

the tip. By spreading this way, a single plant can eventually colonize (take over) huge areas of bare mud. In marsh grass, the rhizomes are soft and fleshy. They lie horizontally at a depth of 2 to 4 inches (5–10 cm). Because they are underground and therefore not photosynthesizing, rhizomes are white or brownish, rather than green. Like the stems aboveground, rhizomes have nodes and internodes. They also have scale leaves growing from the nodes. These papery, nonphotosynthetic leaves are often just sheaths, without any trace of the blade.

Similar to rhizomes are the aboveground creeping stems called stolons or tillers. These stems send out roots at intervals from their nodes, where they touch the soil. Both rhizomes and tillers enable the plant to spread and colonize new ground.

Roots

Marsh grass has two types of roots: finely branched shallow roots and unbranched penetrating roots. The penetrating roots can extend to a depth of 10 feet (3 m) into the soft mud in which the plant grows. The finely branched roots form the fibrous root system, which is where most absorption of water and soil minerals takes place. The young roots have root hairs (tubular extensions of the root's outer cells), which greatly increase the surface area for absorption.

Marsh grass as a land builder

Marsh grass grows on mudflats and tidal estuaries, where repeated flooding by seawater prevents most other plants from growing. Once marsh grass gets established, it can transform the land over a few years from bare mudflats to productive grassland.

The marsh grass's tangle of thick rhizomes holds it firmly in the soft mud. As the mat of grass develops, it absorbs the energy of the waves, protecting the soil and allowing the water to drop any sediment it is carrying. In this way, the soil level can build up quickly—6 feet (1.8 m) in 10 years has been recorded. As the soil level rises above the waterline, the mudflats change from a sea of marsh grass to marshy and then dry meadows where other grasses can grow.

Internal anatomy

As with other green plants, the primary site of photosynthesis in marsh grass is the leaves. Their internal structures are therefore arranged to maximize light absorption and allow in carbon dioxide from the air. Water is carried from the roots to the stems and leaves through xylem tissue. Phloem carries plant sap, with the sugary products of photosynthesis, from the leaves to the rest of the plant.

Inside the stems and leaves

A marsh grass stem, like that of most other grasses, is hollow. Tightly packed epidermis cells protect the outside of the stem. Lines of vascular tissue containing phloem and xylem vessels run through the stem.

The outer surfaces of the leaf consist of layers of cells called the upper and lower epidermis. These cells are arranged regularly in rows that are parallel to the sides of the leaf. The stomata are pores that open and close to allow gas

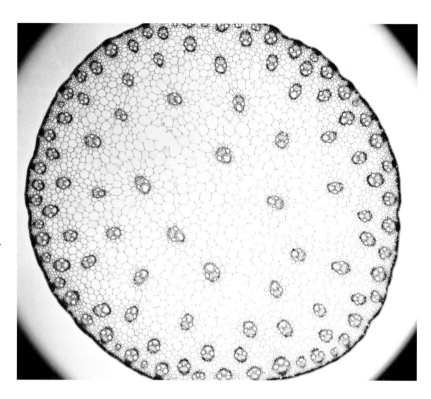

▲ *This cross section of a* Zea mays *(corn) stem shows the irregular distribution of the vascular bundles found in monocot plants. Unlike other grasses,* Zea mays *has a stem that is not hollow.*

exchange between the leaf tissues and the surrounding air. They are found mainly on the underside of the leaf.

Inside the leaf, vascular bundles lie in parallel lines (in contrast to the netlike branching veins of dicots) all long the leaf. The vascular bundles are the plumbing for the leaf, bringing in water and removing the products of photosynthesis. The cells surrounding the vascular bundles, called bundle sheath cells, are responsible for transferring minerals and sugars absorbed or produced in other parts of the plant to their destinations. Bundle sheath tissue can also serve

epidermis

hollow center

Each vascular bundle is made up of **xylem** *and* **phloem** *tissues.*

The **vascular bundles** *are not arranged in neat, concentric rings as in some plants, but are distributed irregularly.*

◀ STEM CROSS SECTION
Most grasses, including marsh grass, have hollow stems. As in all monocots, their vascular bundles are distributed irregularly. Epidermal cells protect the outer surfaces of the stem.

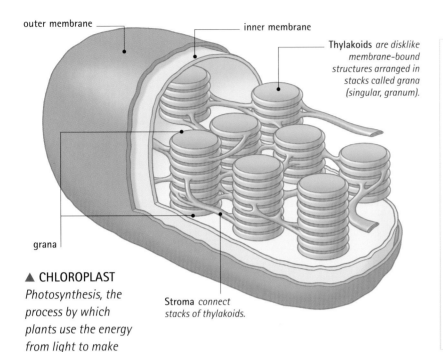

outer membrane

inner membrane

Thylakoids *are disklike membrane-bound structures arranged in stacks called grana (singular, granum).*

grana

▲ CHLOROPLAST

Photosynthesis, the process by which plants use the energy from light to make sugars, occurs in the chloroplasts. Light energy is harvested on the membranes of structures called thylakoids.

Stroma *connect stacks of thylakoids.*

▼ *An electron microscope image of the stacks of grana inside a chloroplast.*

as a temporary store. In marsh grass, the bundle sheath cells contain most of the chloroplasts.

Inside rhizomes and roots

Rhizomes (underground stems) in *Spartina* species have a spongy texture, with a circle of large intercellular spaces separated from one another by radiating plates of cells. Large air spaces in roots and rhizomes help the movement of gases, so the tissues can "breath" in the waterlogged soil in which the plants live.

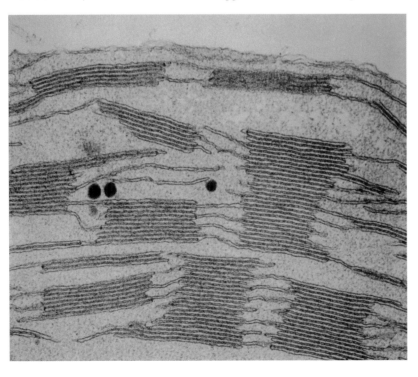

In marsh grass and certain rice species, intercellular spaces in the roots are home to nitrogen-fixing bacteria. These bacteria "fix" nitrogen gas in the air by making it part of a compound that the plants can use. In return, the bacteria feed on carbohydrates that leak from the plant's root cells.

CELL BIOLOGY

C3 and C4 photosynthesis

All green plants can use chlorophyll and the energy of sunlight to build up compounds from carbon dioxide (CO_2) and water. The first product in the chain of chemical reactions involved is a compound (phosphoglycerate) containing three carbon atoms, so the process is described as C3 photosynthesis. In normal C3 plants, this process takes place in the mesophyll cells.

A few plant families, mainly from tropical regions, also have an additional kind of photosynthesis. In C4 photosynthesis, the carbon from CO_2 in the air is captured in a four-carbon molecule (malate or aspartate) within the mesophyll. The four-carbon molecule is then transported to the bundle sheath cells where CO_2 is released, to be used in the normal C3 photosynthesis reaction.

Reproduction

Grasses are successful because they can spread to new areas, quickly colonize an area once they have reached it, and hold their own against other plants. Grasses achieve this success through a combination of sexual and vegetative reproduction.

Grasses reproduce vegetatively using rooting stems called stolons, tillers, and rhizomes. Sexual reproduction usually involves the transfer of pollen from one plant to another. Some grasses self-pollinate, but most avoid this, as the main benefit of sexual reproduction is exchanging genes between plants to create new and potentially better mixtures.

Flower structure

Grasses are usually wind-pollinated, so their flowers lack the showy petals and attractive scents of insect-pollinated plants. Instead, they use breezes to carry pollen to the female flower structures. Because this process relies on chance, grasses produce large quantities of light, dustlike pollen. By contrast, orchids, lilies, and other insect-pollinated monocots produce small amounts of sticky pollen.

The individual grass flowers are small and usually green or brown, and they are grouped on the flower stalk in a structure called an inflorescence. These structures are very varied. They can be closely packed to give a dense spike, as in wheat (*Triticum*), or arranged on long thin stalks to give a loose, feathery structure called a panicle, as in pampas grass (*Cortaderia selloana*).

Spikelets and florets

The basic unit in a grass inflorescence is the spikelet. Each spikelet has at its base a pair of scales called glumes. Sometimes there is a bristle or spine (called an awn) at the tip of one or both glumes. The glumes can be either the same size, or one larger than the other.

The spikelet consists of one or more small flowers (florets). In each floret, the sexual organs are enclosed by two bractlike scales. The lower scale, called the lemma, is usually opaque, greenish, and tougher than the upper

one. In many grasses it has an awn (bristle), either at its tip or arising from its back. The upper scale, called the palea, is usually delicate, thin, and often transparent and silvery.

Most grasses have three stamens per floret. Each stamen has a long, flexible stalk or filament. The anther, containing the copious dustlike pollen, is attached at the tip of the filament by a flexible joint. As a result of this

▲ Spartina anglica *is an invasive hybrid of* Spartina alterniflora *and* Spartina maritima. *Often introduced for land reclamation, in some places it has damaged local ecologies.*

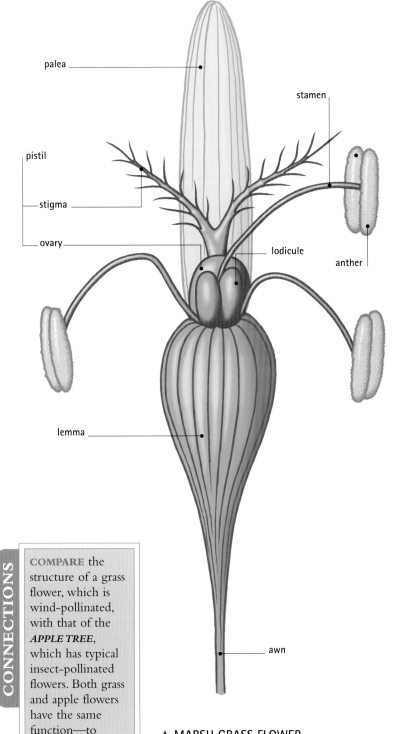

palea

stamen

pistil

stigma

ovary

lodicule

anther

lemma

awn

COMPARE the structure of a grass flower, which is wind-pollinated, with that of the *APPLE TREE*, which has typical insect-pollinated flowers. Both grass and apple flowers have the same function—to ensure that pollen from one plant reaches the female parts of another—but their anatomy reveals the different ways they achieve this.

▲ MARSH GRASS FLOWER
Grass flowers depend on wind for pollination and therefore do not need to have elaborate, showy flowers for attracting insects. Each floret is initially enclosed by two scales, the lemma and the palea, which are split apart by the swelling of the lodicules. The pollen on the anthers is blown by wind to the stigma of another flower, where it fertilizes an egg cell in the ovary.

Useful seeds

Many of the grasses are familiar foods. Wheat, maize, and rice form the staple foods of most of the world's human population. Wheat grain is used to make flour, as livestock feed, and for brewing beer. Most commercially grown wheat is *Triticum aestivum* (bread wheat) or *Triticum durum* (macaroni wheat). The rice we eat is *Oryza sativa* or *Oryz glabberina*. We also use the seeds of many other plants for food and other products. Sunflower (*Helianthus annus*) seeds are pressed to produce sunflower oil. Cotton (*Gossypium*) seeds are covered with long, fine hairs that we spin into threads for making fabric.

arrangement the stamens jiggle in the slightest breeze and disperse the pollen. Also inside the floret are two small bodies called the lodicules. These swell when the floret is mature, forcing the lemma and palea apart and allowing the stamens to emerge to release their pollen.

The ovary (the female part of the flower) is in the center of the floret. It contains a single ovule, which develops into the seed. On top of the ovary, on very short stalks called styles, are two long, feathery parts called stigmata, which are there to catch any pollen that drifts past. The whole female structure is called the pistil.

Seeds

The fruits of grasses are dry and hard. Each fruit is called a grain, or caryopsis, and it contains only a single seed, whose coat is closely attached to the fruit wall. In many other plants, the difference between the fruit and the seed (or seeds) is clear—for example, in tomatoes and plums; and in beans (where the fruit is the whole pod). In grasses, however, because the seed coat is fused to the fruit wall, and there is only one seed per fruit, there is no practical difference between fruit and seed.

The bulk of a grass seed is endosperm—tissue that provides nutrition for the developing plant (embryo) when the seed germinates. In mature seeds, this is hard and contains starch grains. Wheat endosperm is milled to make flour. In

Maize, an unusual grass

Maize, or sweet corn (*Zea mays*), is unusual among grasses. The structure of the inflorescence is unlike any other grass, and indeed nothing very similar grows in the wild. The male and female flowers are separate (unusual for a grass), with the male inflorescence (tassel) at the top of the plant and female inflorescences (ears) lower down. Maize kernels are arranged in rows. There is always an even number of rows, because the spikelets are always paired.

▶ **Maize inflorescence**
Maize flowers grow at the top of a long flower stalk around 9 feet (2.5 m) high. Unlike other grasses, the stem of Zea mays *is solid rather than hollow.*

▼ **GRASS SEED CROSS SECTION**
The seeds of monocots, such as grasses, contain food stored in the endosperm. This is absorbed directly by the cotyledon to power germination.

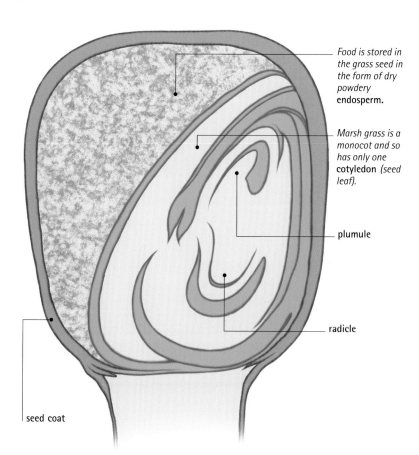

Food is stored in the grass seed in the form of dry powdery **endosperm.**

Marsh grass is a monocot and so has only one **cotyledon** *(seed leaf).*

plumule

radicle

seed coat

sweet corn (maize, *Zea mays*), the embryo is the little spear-shaped structure that sometimes pops out when you bite into a kernel.

Dispersal

Many grass seeds have no specific dispersal mechanism; the seeds just drop out of the inflorescence when they are mature. The grains usually drop while still attached to lemma and palea, which form the husks. In some species, the awns (the spikes on the lemma and palea) twist and change their shape as they absorb water and then dry out again. This twisting and untwisting can push a fallen seed across the soil surface until it falls into a crack. In many grasses, seeds can germinate after passing through an animal's digestive tract. Antelopes, birds, cattle, elephants, and termites have all been found to disperse grass seeds in this way.

ERICA BOWER

FURTHER READING AND RESEARCH
Harrington, H. D. 1997. *How to Identify Grasses and Grasslike Plants: Sedges and Rushes.* Swallow Press: Athens, OH.
Watson, L., and M. J. Dalwitz. 1992. *The Grass Genera of the World.* CABI Publishing: Cambridge, MA.

Muscular system

Muscles help animals move. At some stage in their life cycles, most animals are capable of moving their body from one place to another, perhaps to search for food, a mate, or a suitable place to settle. However, there are many other, more subtle movements that are equally important for the survival of an animal. These are the movements of individual body parts, such as the feeding appendages of barnacles, or internal organs, including the heart and the intestine.

Whatever the tissue or animal species, muscle cells are responsible for all movement in animals that involve more than just individual cells, and they account for almost every movement that can be observed with the unaided eye. Locomotion of individual cells, including unicellular eukaryotes, can be achieved by amoeboid movement or by beating of hairlike structures, such as cilia or flagella, but these will not be discussed here.

Contraction

Muscle cells are able to contract (shorten), and they can do this because they have many copies of two particular proteins: myosin and actin. All cells have versions of these proteins, but in muscle cells they are of particular types, especially abundant, and also organized into strands, or filaments. Myosin and actin filaments can slide against each other, and this is what happens throughout the muscle cell when it is contracting. There are many types of muscle cells; some work individually whereas others are organized into tissues and work together to form different types of muscles.

Muscles that carry out behaviors of the body are usually arranged in pairs or groups that work against each other. In limbs, for example, flexor muscles draw a part of the limb toward the body, and extensors straighten out the limb. To convert the force generated in a muscle to work,

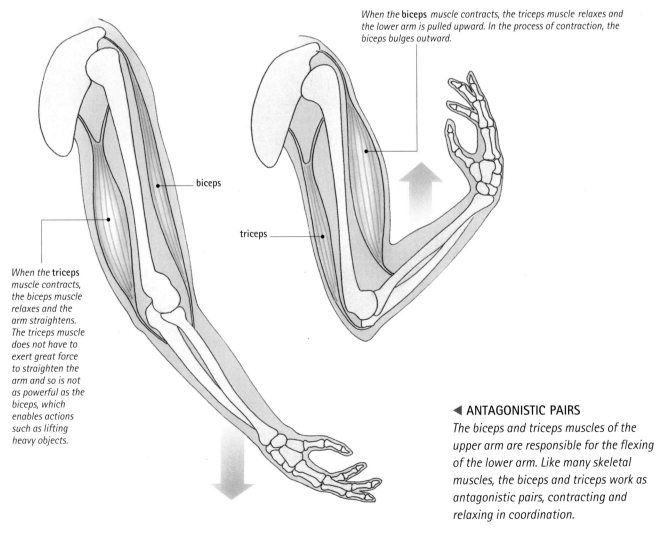

*When the **biceps** muscle contracts, the triceps muscle relaxes and the lower arm is pulled upward. In the process of contraction, the biceps bulges outward.*

biceps

triceps

*When the **triceps** muscle contracts, the biceps muscle relaxes and the arm straightens. The triceps muscle does not have to exert great force to straighten the arm and so is not as powerful as the biceps, which enables actions such as lifting heavy objects.*

◀ ANTAGONISTIC PAIRS
The biceps and triceps muscles of the upper arm are responsible for the flexing of the lower arm. Like many skeletal muscles, the biceps and triceps work as antagonistic pairs, contracting and relaxing in coordination.

tentacles

mouth contracted

▲ **CIRCULAR MUSCLES**
The closing of a sea anemone's slitlike mouth is controlled by rings of circular muscles.

▶ **SECTION THROUGH A SEA ANEMONE**
A sea anemone does not have a hard skeleton. Instead, it has a hydrostatic skeleton: muscles contract against water held in compression by the anemone's tissues. The water is thus forced into other parts of its body, causing expansion or contraction.

The column is shortened by the contraction of longitudinal muscles.

The column is lengthened by the contraction of circular muscles.

tentacles

Retractor muscles of the septa.

◀ **EXTENSION AND CONTRACTION**
The sea anemone uses longitudinal muscles that run perpendicular to the base and circular muscles parallel to the base to extend and contract the column, withdraw and extend the tentacles, and open and close the mouth.

the muscle must pull or press on a rigid structure of some kind. This structure could be an endoskeleton, such as our bones; or an exoskeleton, such as the cuticle of an insect or the shell of a clam.

Sometimes, the function of the muscles is to lengthen or shorten a structure that is not supported by a rigid skeleton. Muscles in sea anemones, for example, enable these soft-bodied animals to stretch out to catch food or retract to a small budlike shape. Layers of longitudinal and circular muscles work together to achieve such movements; contraction of the longitudinal muscles shortens the sea anemone, whereas contraction of the circular muscles

HOW MUSCLES WORK Muscles are generally either smooth or striated. Voluntary muscles are striated, and involuntary muscles are smooth. *See pages 756–761.*

SKELETAL MUSCLE Used for functions such as locomotion, skeletal muscles are attached to skeletal structures such as the bones of vertebrates or the exoskeletons of some invertebrates. *See pages 762–765.*

CARDIAC MUSCLE This type of striated muscle has only one nucleus per muscle fiber. Contraction of the heart muscle is coordinated by the nerve bundles of the sinoatrial and atrioventricular nodes. *See pages 766–767.*

SMOOTH MUSCLE Smooth muscle lacks the obvious internal arrangement of striated muscles and are responsible for involuntary functions such as movement of food through the intestine. *See pages 768–769.*

FEATURED SYSTEMS

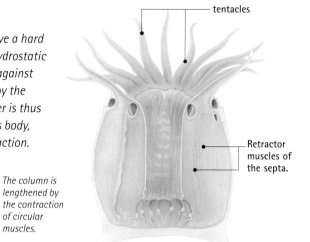

▲ *A network of tiny blood capillaries (red) supplies the muscles (brown) with the oxygen and nutrients they use to produce energy by the process of respiration.*

lengthens it. However, even in the case of the body movement of a sea anemone, the muscles are working against a "skeleton." This skeleton is the water that is retained in the hollow center of the animal. Because an enclosed volume of water does not compress if pressure is applied to it, the water inside a sea anemone functions as a hydrostatic skeleton against which the force of the muscles can be converted into work. In principle, this is also what happens in our heart muscle, but here the contraction of the heart forces the blood to pass into the circulatory system.

How muscles work

COMPARE the arrangement of cells in smooth and striated muscles with the arrangement of cells in other tissues, such as the alimentary canal of the *DIGESTIVE AND EXCRETORY SYSTEMS.*

Muscles can be generally classified as either striated or smooth. In vertebrates, including humans, smooth muscles are responsible for many of the body functions we are normally not aware of, including regulating blood flow through blood vessels, expelling secretions from various glands, and moving food along the intestine. Striated muscles include both the heart muscle and the skeletal muscles; the latter move the bones of our skeleton in relation to each other. The term *striated* comes from the banded appearance of these muscles when viewed under a microscope. The bands are caused by the ordered organization of contractile myosin and actin filaments in striated muscle fibers.

Muscle structure

Structures within skeletal muscles were first studied using light microscopy and have therefore been given names according to their appearance. The primary repeated structural units of striated muscles are referred to as sarcomeres. Neighboring sarcomeres are divided by a prominent dense line of tissue called the Z disk. The so-called A bands span over the region of a sarcomere where the contractile myosin filaments are located. Each myosin filament is surrounded by thin actin filaments, which are anchored to the Z disk.

When a muscle is relaxed, the myosin filaments occupy the middle part of the sarcomere, leaving a region containing only actin filaments and the Z disk. This region forms a light band stretching from the end of the myosin filaments in one sarcomere to the beginning of the myosin filaments in the next. This lighter region is named the I band. The actin filaments are made up of globular G-actin molecules joined together like pearls on a string to form a double helix. In the vertebrate skeletal muscle, each myosin filament is surrounded by six actin filaments. The myosin

▼ MUSCLE TYPES
Muscles are divided into two main types: striated muscles, which make up skeletal muscles; and smooth muscles in, for example, the walls of the intestine, arteries, and veins.

▼ CARDIAC MUSCLE
In cardiac muscle, individual cells are branched and connected to one another by junctions called intercalated disks.

Striated muscle fibers

A bands

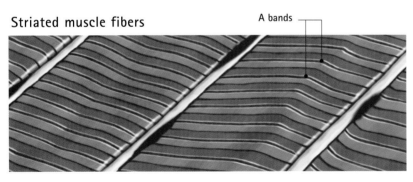

Smooth muscle fibers

nuclei

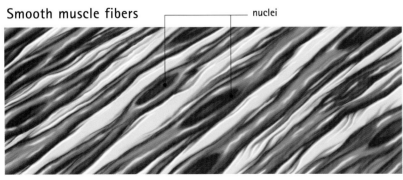

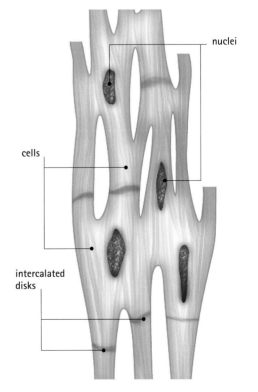

nuclei

cells

intercalated disks

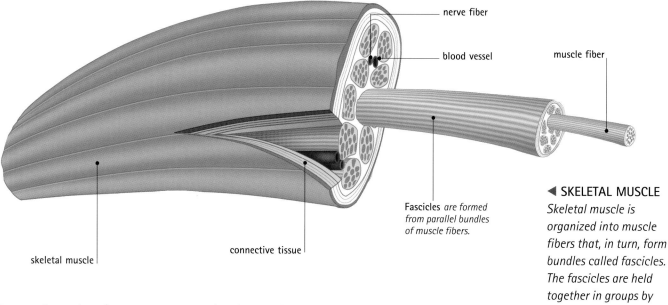

nerve fiber

blood vessel

muscle fiber

skeletal muscle

connective tissue

Fascicles *are formed from parallel bundles of muscle fibers.*

◀ SKELETAL MUSCLE
Skeletal muscle is organized into muscle fibers that, in turn, form bundles called fascicles. The fascicles are held together in groups by connective tissue.

filament is made of many myosin molecules bundled together. A myosin molecule contains a "head region," a "neck region," and a "tail region." It is made up of two long proteins and several small proteins located in the head region. The head of the myosin molecule contains the machinery that binds to actin and causes the contraction.

At the junction between the A and the I bands of the muscle fiber, there are numerous minute canals that tunnel from the contractile membrane—or sarcolemma—surrounding the muscle fiber to the interior of the muscle fiber. These canals are called transverse tubules (or T-tubules), which are continuous with the environment around the muscle fiber, make contact with every myofibril. The muscle fibers in striated muscles also have a well-developed sarcoplasmic reticulum, which is a baglike membranous structure corresponding to the endoplasmic reticulum of most other cells. The sarcoplasmic reticulum is located within the muscle fiber, and it stretches over each sarcomere. The sarcoplasmic reticulum is spread out over the myofibrils. Close to the T-tubules, it enlarges into terminal cisterns.

Cell biology of muscle contraction

It was first believed that the myosin and actin filaments became shorter during contraction of the muscle fiber, but that is not the case. Instead, the myosin and actin filaments slide in relation to each other. This makes the sarcomere shorter, but the length of the filaments themselves remains unchanged. What is happening is that the myosin heads are "walking" up the actin filaments, pulling them in line with one another. Two scientists, Hugh Huxley from the Massachusetts Institute of Technology and Andrew Huxley from Woods Hole Marine Biological Institute, introduced this sliding-filament theory independently in 1954. The theory has been revised many times, leading up to the current model.

CLOSE-UP

Muscle organization

Vertebrate skeletal muscles are composed of many bundles of long muscle fibers, which are the actual muscle cells. These bundles of muscle fibers are called fascicles, and within any fascicle all muscle fibers run in parallel. Each muscle fiber has many nuclei, because it forms through fusion of several immature muscle cells during development. Each muscle fiber contains numerous parallel elements called myofibrils. The myofibrils, in turn, consist of repeated units called sarcomeres. It is the sarcomere that is the functional contractile unit of the muscle.

FASCICLE

sarcolemma

myofibrils

▼ SKELETAL MUSCLE DETAIL
The illustrations show progressive levels of detail from a fascicle to the molecular level. A fascicle contains many myofibrils, which are formed from alternating regions of thick and thin filaments that slide over one another, resulting in muscle contraction.

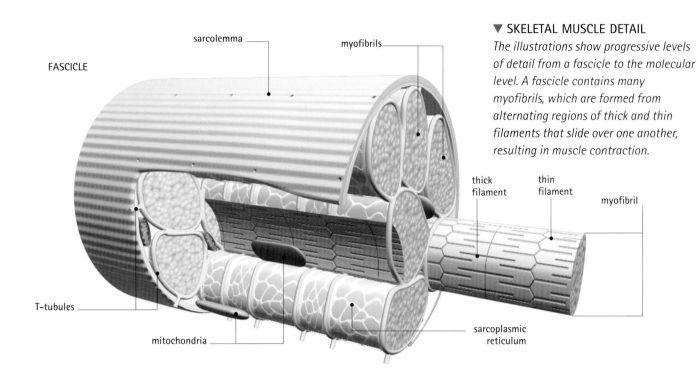

thick filament

thin filament

myofibril

T-tubules

mitochondria

sarcoplasmic reticulum

MYOFIBRIL

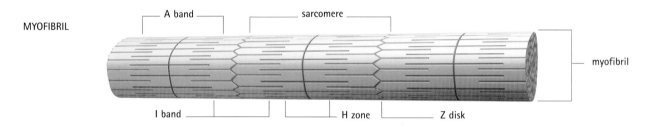

A band

sarcomere

myofibril

I band

H zone

Z disk

Myosin heads "walk" along actin chains, contracting muscles.

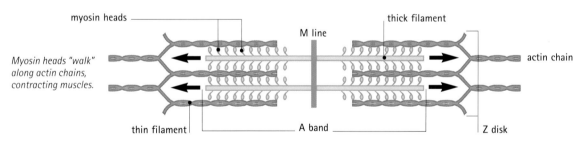

myosin heads

M line

thick filament

actin chain

thin filament

A band

Z disk

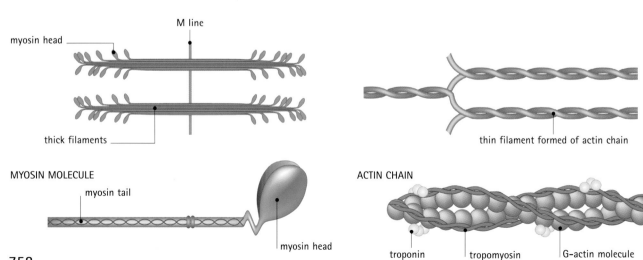

M line

myosin head

thick filaments

thin filament formed of actin chain

MYOSIN MOLECULE

myosin tail

myosin head

ACTIN CHAIN

troponin

tropomyosin

G-actin molecule

The sliding of actin and myosin filaments relative to each other leads to shortening of the sarcomere and contraction of the muscle. This process requires energy, and the energy is provided in chemical form by adenosine triphosphate (ATP), which is an energy-storage molecule found in all living cells. The ATP binds to the myosin head so that when the contraction cycle starts, the myosin head already has one molecule of ATP bound. The myosin head then binds to the actin filament; and with the energy released from the breakdown of the ATP to adenosine diphosphate (ADP), the myosin head turns from a 90-degree angle relative to actin to a 45-degree angle. Owing to this power stroke the myosin and actin filaments slide against each other. This is where the force of the contraction is created. During the power stroke, the ADP is released, leaving space for a new ATP to bind to the myosin head. Only once the ADP is replaced with a new ATP can the cross bridge between the myosin and actin filaments be broken and a new cycle be ready to start. In each cycle the myosin "walks" a little farther up the actin filament. The distance that the filaments slide against each other during the turning of one myosin head is tiny, but each sarcomere has many myosin heads that repeatedly go through the same cycle, and in a muscle fiber there are many sarcomeres stacked one after another. This makes the accumulated movement of myosin heads sufficient to shorten the muscle and move the body.

Role of calcium ions

In addition to energy in the form of ATP, filament sliding and muscle contraction need calcium ions (Ca^{2+}). An increased Ca^{2+} concentration in the cytosol (the semifluid substance that fills most of the space in a cell, except the organelles), or sarcoplasm, as it is called in muscle fibers, is the trigger that causes myosin heads to bind to actin filaments. Ca^{2+} starts the formation of myosin–actin cross bridges by binding to a protein associated with the actin filament. In the groove between the two intertwined actin strands, there is a long threadlike protein called tropomyosin. At distances of 400 nanometers (0.0004 mm) apart, representing a half turn of the actin

IN FOCUS

Rigor mortis

A little while after an animal dies, the body goes through a period during which it is stiff for several hours. This stiffness is called rigor mortis. After death, adenosine triphosphate, ATP, which is used for muscle contraction, is rapidly broken down throughout the body and disappears from the muscles. As ATP is needed for the myosin heads to release their bonds to the actin filaments, the muscles become fixed and rigid. The body remains in rigor mortis until the muscle decomposes to the extent that the myosin and actin filaments are destroyed.

double helix, there is a protein complex called troponin complex. The troponin complex is made of three subunits. One of these subunits binds actin, one binds tropomyosin, and one (the C-subunit) binds Ca^{2+}.

▼ To power the contraction of the muscles in these runners' legs, energy is released from large amounts of adenosine triphosphate, ATP. This release of energy enables myosin filaments to "walk" along actin chains in the muscle fibers, thus shortening the length of the muscles.

▼ CONTRACTION

Muscle contraction can be divided into six phases. (1) The myosin head is tightly bound to a myosin binding site at an angle of 45 degrees to the actin chain (labeled site a). (2) An ATP molecule binds with the nucleotide binding site, causing the myosin head to detach from the myosin binding site. (3) ATP is converted into adenosine diphosphate (ADP) and a phosphate. This action releases energy. (4) The myosin head rotates backward and attaches loosely to another actin site (labeled b) at an angle 90 degrees to the actin chain. (5) The inorganic phosphate is released, causing the myosin head to swing forward, thus pushing the actin chain forward. (6) The ADP is released, and the myosin head is once more bound tightly to the actin binding site.

When the muscle is relaxed, the tropomyosin covers the myosin-binding site on actin, preventing the myosin head from binding to it. If the Ca^{2+} concentration in the sarcoplasm is increased (to about 10 μm), four Ca^{2+} ions bind to the C-subunit of the troponin complex. Binding of Ca^{2+} to troponin C causes the whole troponin complex to retract, and in the process it pulls the tropomyosin molecule away from the myosin-binding site on each actin subunit. Now the myosin-binding site is exposed, and the myosin head can bind to the site, causing the sliding of filaments and, hence, contraction of the muscle to commence.

In a normal muscle fiber, the contraction will continue as long as sufficient ATP and Ca^{2+} are present close to the actin filaments. To end the contraction, Ca^{2+} must be removed from the sarcoplasm. To achieve this, the membrane of the sarcoplasmic reticulum is lined with Ca^{2+} pumps that move Ca^{2+} from the sarcoplasm back to the inside of the sarcoplasmic reticulum, which functions as a Ca^{2+} reservoir. Inside the sarcoplasmic reticulum Ca^{2+} binds to a special protein, which increases the reticulum's storage capacity of Ca^{2+}. The Ca^{2+} pumps are sufficient to keep the concentration of free Ca^{2+} in the sarcoplasm extremely low when the muscle is relaxed.

Action potentials

A contraction in a skeletal muscle starts with a series of electrical pulses (action potentials, or nerve impulses) traveling down a nerve to reach the individual muscle fibers. Each nerve cell, or nerve fiber, controls the contraction of several muscle fibers. Muscle fibers connected to the same nerve fiber contract and relax together; they are therefore called a motor unit. The nerve fiber ends in a nerve terminal that is very close to—but does not make direct contact with—the muscle fiber. This point of communication between nerve fiber and muscle fiber is a type of synapse called a neuromuscular junction.

In vertebrates, when the action potentials that have been traveling down the nerve fiber reach this terminal a chemical called acetylcholine is released from the nerve terminal. This neurotransmitter diffuses along the short distance from the nerve terminal to the sarcolemma of the muscle fiber. Embedded in the sarcolemma is a protein that binds the acetylcholine and functions as a receptor,

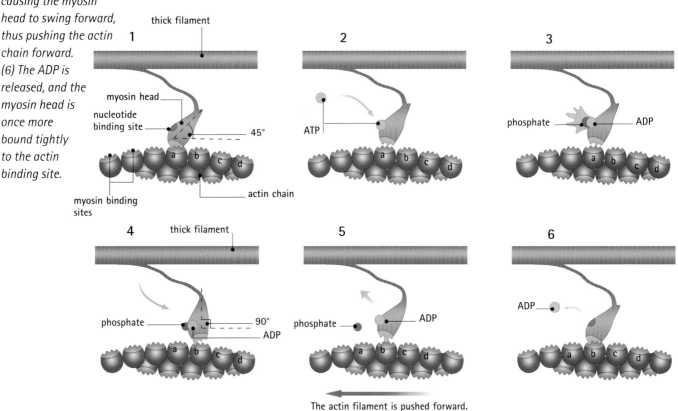

The actin filament is pushed forward.

mediating the signal to the muscle. Binding of acetylcholine to the acetylcholine receptor triggers the opening of an ion channel in this receptor protein. The channel allows positively charged sodium ions (Na$^+$) to enter into the muscle fiber. As a result of the inward flow of Na$^+$ ions, the inside of the muscle fiber becomes more positively charged. This charge triggers action potentials along the sarcolemma of the muscle fiber. When the action potential reaches a T-tubule, it follows the T-tubule into the interior of the muscle fiber.

In the membrane of the T-tubule, close to the terminal cisterns of the sarcoplasmic reticulum, there are voltage-sensitive proteins (dihydropyridine receptors) that change shape in response to the incoming action potentials. These voltage-sensitive proteins make direct contact with Ca^{2+} release channels (ryanodine receptors) located in the membrane of the sarcoplasmic reticulum. When stimulated by an action potential, the voltage-sensitive proteins in the T-tubule membrane force the Ca^{2+} release channels to open a pore through which Ca^{2+} is able to escape from the sarcoplasmic reticulum. As a consequence, the Ca^{2+} concentration of the sarcoplasm increases. This increase, in turn, results in binding of Ca^{2+} to troponin, cross-bridge formation between the myosin and actin filaments, active filament sliding, shortening of the sarcomere, and contraction of the muscle.

Energy for muscle work

Muscle cells need energy to carry out work; and, as noted above, this is provided in the form of energy-rich ATP. In animals ATP is produced primarily through the reaction between breakdown products of foodstuffs and molecular oxygen. Some muscle cells, such as those in the heart and in postural muscles, need an ample supply of oxygen to function. Other muscle cells can operate without oxygen for some time. During this time, these muscle cells make ATP from reactions that do not require oxygen. In this process, the body builds up an "oxygen debt," which later has to be repaid by a higher-than-normal oxygen uptake after the exercise.

Muscle cells that mostly rely on constant oxygen supply are typically reddish brown in appearance. This is for two reasons: first, they

contain many mitochondria, which are sausage-shaped organelles responsible for oxygen-dependent ATP production. Second, these muscle cells are typically rich in myoglobin, which is an oxygen-binding protein similar to hemoglobin. Myoglobin has a higher affinity for oxygen than does hemoglobin and is therefore able to "extract" oxygen from the blood and capture it for the muscle tissue. What we often refer to as red meat contains a high proportion of muscle fibers that require a rich oxygen supply; white meat has relatively more muscle fibers that can work without oxygen. In general, red muscle cells contract slowly but also fatigue slowly; white muscle cells contract quickly but also fatigue quickly.

▼ *Nerve axons (the dark lines) connect to individual skeletal muscle fibers (the broad parallel bands) across a network of neuromuscular junctions, or motor end plates. The black dots are synapses. Magnified 400 times.*

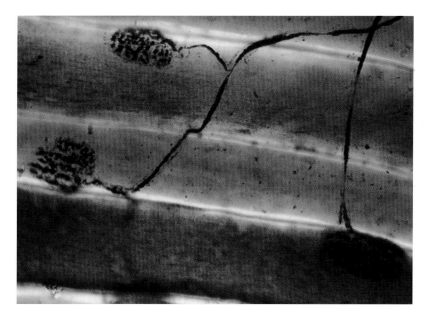

Skeletal muscle

CONNECTIONS

COMPARE
the form and arrangement of skeletal muscles in a fish, such as a *TROUT,* with their form and arrangement in a marine mammal, such as a **DOLPHIN** or **GRAY WHALE.**

Although the basic mechanisms of most skeletal muscles are the same, differences in their structure and arrangement allow them to perform different jobs. Skeletal muscles in animals perform an astonishing range of tasks, including running, flying, sound production, and, in some fish, generation of electricity. Skeletal muscle attaches to bones in vertebrates and to the hard exoskeleton of many invertebrates. The locomotor muscles of an earthworm are adapted to create waves of contractions that spread along the body in a highly synchronized fashion. These contractions alternate between circular and longitudinal muscles to produce the waves of elongation and contraction that move the earthworm forward through the soil. The muscles in the claws of many decapod crustaceans, such as crabs, lobsters, and crayfish, are uniquely adapted to generate an incredibly strong force. These muscles are often short with a large cross-sectional area, and the fibers are arranged at an angle to the direction of the pull. Such an arrangement produces a short range of motion but great power; it also allows the claw muscle to contract without bulging. This is an advantage because the space inside the claw is limited by the rigid exoskeleton.

Some muscles can contract and relax extremely quickly. For example, the muscles that operate the wings of some dipteran insects such as hoverflies can contract and relax several hundred times per second, providing the characteristic high-frequency wing beat of this animal group. Interestingly, this rate is much faster than that of action potentials in the nerve fibers that control the flight muscles. Researchers have found that once a cycle of alternate contractions in elevator and depressor wing muscles has been initiated, this cycle is propagated without the need for an action potential for every contraction. In these so-called asynchronous flight muscles the action potentials serve as an on and off switch, and the wings continue to beat so long as action potentials keep coming to the muscles.

▶ **INSECT WINGS**
An insect is an invertebrate, so it does not have an internal skeleton. Instead, an insect's hard exoskeleton serves as an anchor for muscles. The upstroke of an insect's wing is powered by the contraction of dorsoventral muscles. These muscles pull on the tergite (the top of the insect's thorax), causing the tergite to flex downward and the wings to flex upward. During the downstroke, the reverse happens. The dorsoventral muscles relax, and the basilar and dorsal longitudinal muscles contract, causing the tergite to be pushed upward and the wings to flex downward.

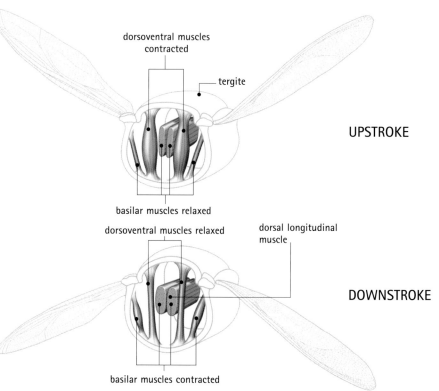

dorsoventral muscles contracted

tergite

UPSTROKE

basilar muscles relaxed

dorsoventral muscles relaxed

dorsal longitudinal muscle

DOWNSTROKE

basilar muscles contracted

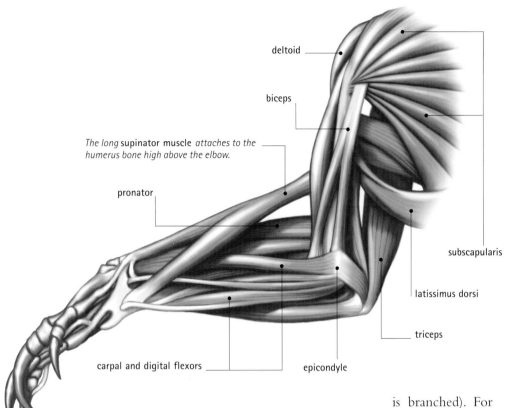

deltoid

biceps

The long **supinator muscle** *attaches to the humerus bone high above the elbow.*

pronator

carpal and digital flexors

epicondyle

triceps

latissimus dorsi

subscapularis

◄ ANTEATER ARM MUSCLES
The skeletal muscles of a giant anteater's forelimbs are particularly well developed. The anteater uses the strength these muscles provide to claw its way into rock-hard termite mounds.

As we have seen, muscles are adapted for different tasks. Some of these demand speed, others strength, and yet others endurance. One single type of muscle would clearly not be able to carry out such a variety of functions. Evolution has led to specialization of the skeletal muscles in different species as well as among the various muscles within an individual animal.

Arrangement of fascicles

Skeletal muscles are attached to the skeleton or other structures by tendons, and they generate movement by applying a force on the tendons, which in turn produces a pull on the attached bone. As noted previously, muscle fibers in a muscle are arranged in bundles, called fascicles. Within each fascicle, all muscle fibers are parallel to each other. However, as in the case of the crab claw muscle discussed above, fascicles are not always parallel. In fact, there are at least five different patterns of fascicle arrangements, each producing a specific property and usage of the muscle: parallel, fusiform, circular, triangular, unipennate (with muscle fibers on the same side of a tendon), bipennate (with muscle fibers on both sides of a tendon, and multipennate (where the tendon

is branched). For example, many muscles in our limbs have either a parallel or a fusiform (cigar-shaped, with the fascicles nearly parallel) arrangement. Sphincter muscles that close an opening, such as the mouth, have a circular fascicle arrangement, in which the fascicles encircle the opening. The closure muscle in the crab claw is bipennate. The fascicles are arranged in an angle on either side of centrally located tendons.

Fiber types

Fascicle arrangement is one way by which muscles can be tailored for a specific task. Another major difference between muscles used for different purposes is the properties of the individual muscle fibers. Most muscles contain more than one type of muscle fiber. In vertebrates, there are four main types of skeletal muscle fibers, each with a specific property. These four types of skeletal muscle fibers can be divided into two principal groups: tonic and phasic. Tonic muscle fibers occur in muscles that maintain the posture of animals such as amphibians, reptiles, and birds. They are also present in the muscles that move the eyeball in its socket. Tonic fibers contract slowly, and each fiber is able to produce a graded contraction. In contrast, a phasic fiber responds to a nerve

763

▶ LOWER FACIAL MUSCLES

The orbicularis oris muscle surrounds the mouth and is a type of sphincter muscle: the fascicles are arranged in a circle.

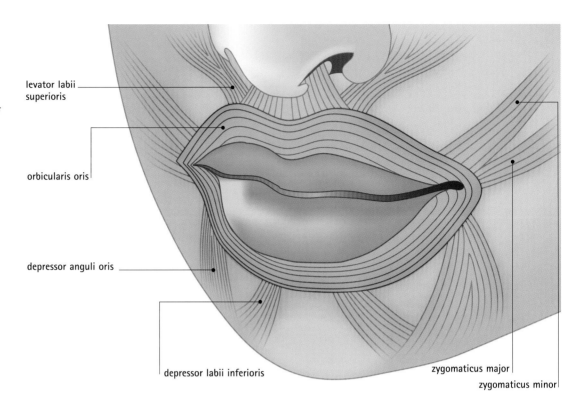

levator labii superioris

orbicularis oris

depressor anguli oris

depressor labii inferioris

zygomaticus major

zygomaticus minor

impulse in an all-or-nothing fashion: that is, either the fiber contracts with maximal force or it does not contract at all.

There are three types of phasic fibers: type I is slow oxidative (SO); type IIa is fast oxidative-glycolytic (FOG); and type IIb is fast glycolytic (FG). SO fibers are found in mammalian muscles that maintain body posture. They have a dark red appearance because they contain large amounts of the respiratory pigment myoglobin and many mitochondria. These enable the cells to generate ATP by aerobic processes. SO fibers contract slowly and also relax slowly.

FOG fibers contain fewer mitochondria and less myoglobin than SO fibers. That is because FOG fibers generate some ATP by anaerobic processes and some through aerobic respiration. They contract quickly and they also fatigue relatively slowly. FOG fibers are specialized for repetitive movements, such as walking and running, and they are especially abundant in the flight muscles of migratory birds such as ducks, seabirds, and warblers.

FG fibers contract very quickly, but they are also very quick to fatigue. They contain little myoglobin and few mitochondria, and they therefore have a much lighter appearance than the other types of fibers. In humans, these fibers are used for "burst activities," such as throwing a ball. Typically, skeletal muscles have a mixture of phasic fiber types, of which about half are SO fibers. The proportions vary depending on the role of the muscle.

During exercise, fast glycolytic fibers are engaged first, and the oxidative types of fiber take over for more prolonged or strenuous activity. In contrast to the mixture of fiber types in most vertebrate skeletal muscles, the muscle fiber types of swim muscles in many fish are anatomically separated from each other. In fish such as mackerel and tuna, which for most of the time cruise at relatively low

---

CLOSE-UP

Force versus speed

When muscles contract, they pull two skeletal structures closer together. Often a muscle on the one side of a joint is pulling on a bone on the other side of the joint. The bone that is moved functions as a lever; and for levers there is a trade-off between the speed and force of the movement. Greater speed can be generated at the expense of reduced force if the insertion point of the muscle's attachment to the bone is close to the joint. Conversely, the strength of the movement can become greater if the insertion point is farther away from the joint, but in this case the potential for speed in the movement is reduced.

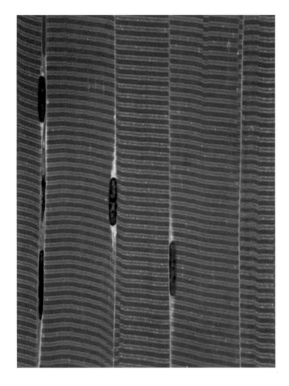

▲ *Typically, around half of an animal's skeletal muscle is made of SO fibers. Skeletal muscle has distinctive striations, and each cell has many nuclei. The dark stripes are A bands; between them are the I bands; and in the middle of the I bands are the Z lines.*

IN FOCUS

Mating call

Many species of bony (teleost) fish communicate with sounds. The mating call of the male oyster toadfish is said to sound like a steamboat whistle and is so loud that it can be heard from above the surface of the water. The sound is produced by muscles drumming on a gas-filled balloonlike structure, called the swim bladder. The teleost swim bladder is best known for functioning to keep a fish neutrally buoyant in the water; but it also has other roles, in hearing and sound production. To produce sound the muscles drum against the swim bladder at a very fast rate. In oyster toadfish the sound-producing muscles are able to contract and relax 200 times per second.

speeds, SO fibers are located in a band just under the skin, roughly along the lateral line. FG fibers, which in many species dominate the muscle mass, are located more deeply. Whereas the SO fibers are used for cruising, the FG fibers are recruited for fast swimming and for the escape response. The latter is the few forceful fin strokes that a fish makes to get away when startled. In squid, there is a similar anatomical differentiation between oxidative and glycolytic swim muscles.

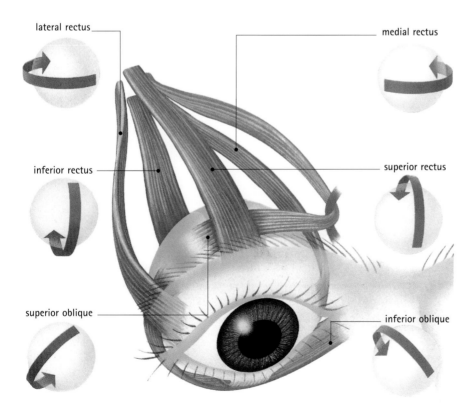

◀ EYE MUSCLES
The muscles around the eyes are made of tonic fibers. The contraction and relaxation of tonic fibers is much slower than that of the more common phasic fibers. Tonic fibers use less energy than phasic fibers and are thus useful where sustained contraction is necessary, such as keeping eyes open and moving throughout the day. Human eyes are moved by six main muscles.

Cardiac muscle

COMPARE the form of a vertebrate heart, such as that of a *HUMAN* or an *ELEPHANT* with that of an invertebrate, such as a *DRAGONFLY* or a *GIANT CLAM*.

The vertebrate heart is made of cardiac muscle. Like skeletal muscle, cardiac muscle appears striated under a microscope. This is because it has regularly arranged sarcomeres, complete with the same bands and Z disks as the skeletal fibers. There are, however, notable differences between skeletal and cardiac muscles. While skeletal muscle fibers have the structure of long and parallel "cables" of fused cells, cardiac muscle cells are much shorter and have only one nucleus each.

A distinguishing feature of cardiac muscle cells is that they branch and are connected to neighboring muscle cells by intercalated disks. The intercalated disks contain proteins that keep the cells tightly joined at junctions called desmosomes as well as proteins that form pores (called gap junctions) between neighboring cells. The gap junctions allow electric currents to be spread between muscle cells.

In addition to the muscle cells that are able to contract, the vertebrate heart has a special electrical conduit system that is made from modified muscle cells that cannot contract. The conducting fibers include separate bundles of conducting fibers that spontaneously generate rhythmic electric impulses. This is a feature that distinguishes vertebrate cardiac muscle from skeletal muscle, which requires nervous stimulation to contract. Another important difference between these two types of striated muscle is that the duration of each electrical impulse spread over the membranes of the muscle cells is 10 to 15

▶ HEARTBEAT
REGULATION
Changes in the rate and strength of the heartbeat are regulated by signals from the autonomic nervous system (ANS), which is not under our conscious control. Information from sensory nerves travels to the regulatory center in the spinal cord. The heartbeat is then adjusted by two different branches of the ANS: the parasympathetic branch and the sympathetic branch.

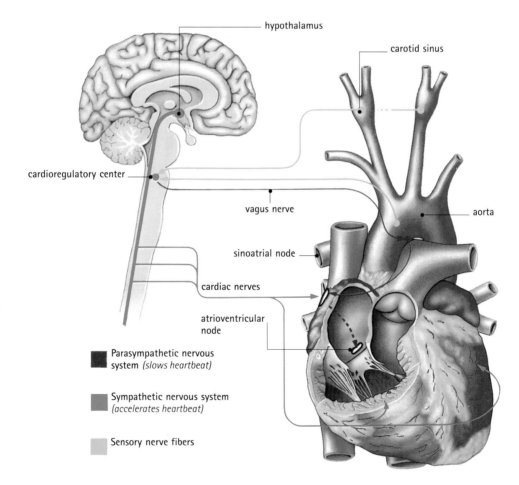

hypothalamus

carotid sinus

cardioregulatory center

vagus nerve

aorta

sinoatrial node

cardiac nerves

atrioventricular node

■ Parasympathetic nervous system *(slows heartbeat)*

■ Sympathetic nervous system *(accelerates heartbeat)*

■ Sensory nerve fibers

times longer in the cardiac muscle than in skeletal muscle. The result is that the contraction in each cardiac muscle cell is also of much longer duration than the contraction in the skeletal muscle fibers.

Nerve nodes

One of the two bundles of conducting fibers that can spontaneously discharge rhythmic electrical impulses is called the sinoatrial (SA) node. The SA node functions as the pacemaker for the heart and initiates the rhythmic contractions of the heart muscle of all mammals. In mammals, the SA node is located in the wall of the right atrium, and from there the electrical impulses are spread through pathways of conducting fibers over the two atria, resulting in the contraction of the right and left atria in succession.

The conduction pathways lead the electrical impulse to the second fiber bundle, called the atrioventricular (AV) node, which connects the atria to the ventricles electrically. From here, the conducting fibers direct the electrical signal along either side of the wall that separates the two ventricles down to the bottom of the ventricles. The conduction fibers then spread the electrical impulse back up along the walls of the ventricles. When each impulse arrives, the muscle fibers in its path contract, resulting in the squeezing of blood out of the respective heart chamber. The electrical impulse starts in the right atrium, and this is, consequently, the first chamber to contract quickly—followed by the left atrium and then, after a short delay, the ventricles. The brief pause between atrial and ventricular contractions is caused by a built-in delay in electrical conduction from the atria to the ventricles across the AV node. Because blood flows from the atria to the ventricles, this delay permits the atria to complete the filling of the ventricles before the latter contract to eject the blood out from the heart.

As a result of the spontaneous electric discharges by the pacemaker cells, vertebrate hearts contract rhythmically without the need for any stimulation by nerves. Even if the heart is surgically removed from the body, it will continue to beat for some time. If sufficiently supplied with oxygen and nutrients, hearts from many lower vertebrates can continue to

beat for hours on their own. Hearts that generate their own rhythmic contractions are called myogenic. Vertebrate hearts do receive signals from nerves, but rather than initiating contractions (as in skeletal muscle) they modulate the rate and strength of the heartbeat. Mammalian hearts are supplied with nerves by nerve fibers belonging to two separate branches of the autonomic nervous system (the part of the nervous system that controls our bodies without our conscious involvement): the parasympathetic branch and the sympathetic branch.

Neurogenic heart

Crayfish, lobsters, and other decapods have neurogenic hearts. A heart is neurogenic if the stimulation to initiate each heartbeat comes from nervous tissue. In the lobster, each muscle cell is connected to a nerve fiber. The muscle cell contracts when it is stimulated by a nerve impulse. The rhythmic pattern of nerve impulses that directs the lobster heart is generated by a cardiac ganglion, which is an aggregation of nine nerve cells at the top of the heart. Five of these nerve cells make connections with the different heart cells, and the other four function as pacemakers that spontaneously generate a pattern of impulses. The pacemaker cells make contact with the nerve cells that supply the heart, and these direct contraction in individual heart cells.

▼ *This colored electron micrograph image of cardiac muscle clearly shows the circular mitochondria between pink muscle fibers. The thick transverse bands of the fibers are Z lines, regions with a denser arrangement of actin muscle filaments.*

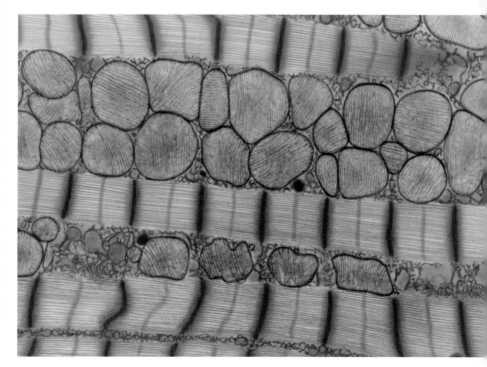

Smooth muscle

Smooth muscle cells are distinguished from striated muscle in that they lack internal arrangement into ordered sarcomeres; they are therefore not striated in appearance when viewed under a microscope. In vertebrates, smooth muscles are important for many of the functions in the body that are normally not under voluntary control. These include most muscles in the digestive system, the reproductive systems, the blood vessels, and the respiratory tract. Smooth muscle is also widespread in invertebrates, but invertebrate smooth muscle does not fall into the two functional groups present in vertebrates: single-unit and multiunit smooth muscles. However, even vertebrate smooth muscle is diverse and does not lend itself well to categorization.

Defining characteristics

Like cardiac muscle, but in contrast to skeletal muscle, each smooth muscle cell is an individual cell with one nucleus. Although sarcomeres are absent in smooth muscle, the contractile machinery is similar in smooth and striated muscle. Smooth muscle cells have myosin and actin filaments and many of the other proteins involved in contraction of striated muscle. Furthermore, as in striated muscle, contraction is achieved by the sliding of myosin and actin

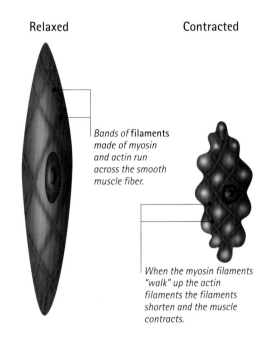

Relaxed   Contracted

Bands of filaments made of myosin and actin run across the smooth muscle fiber.

When the myosin filaments "walk" up the actin filaments the filaments shorten and the muscle contracts.

▲ SMOOTH MUSCLE FIBER
Smooth muscle fibers have one nucleus and are long when relaxed. When they contract, the crisscross arrangement of actin and myosin filaments causes the fiber to become compressed and globular.

filaments past each other as the myosin heads "climb" along the actin filaments. Instead of forming sarcomeres, the myosin and actin filaments of smooth muscle cells are organized into bundles close to the plasma membrane. The actin filaments are attached to dense bodies within the cytoplasm and to attachment plaques at the inside of the plasma membrane. These attachment sites are the functional equivalents of the Z disks in striated muscle.

Role of calcium ions

Smooth muscle cells have poorly developed sarcoplasmic reticuli and no T-tubules. As is the case with striated muscle, an increase in the Ca^{2+} concentration of the cytosol triggers contraction in smooth muscle. However, in smooth muscle much of the increase in cytosolic Ca^{2+} comes from the opening of Ca^{2+} channels in the plasma membrane, allowing Ca^{2+} to flow into the cell from outside. This flow results in a slow rise in the Ca^{2+} concentration of the cytosol

Bivalve catch muscle

In bivalves, such as clams and mussels, the two shells are kept closed by adductor muscles that are attached to either shell. When the creature is attacked by a predator, or is out of water during low tide, the adductor muscle keeps the two halves tightly closed, often for long periods of time. A regular muscle would tend to get tired and use up valuable energy conducting such a task. As a solution to these problems, the adductor muscles of clams and mussels are often a composite between striated fibers, which can deliver speed; and smooth muscle, which can provide endurance. In addition, the smooth muscle is capable of "locking" in a contracted state. No extra energy is required to maintain the muscle in the "locked" contracted position. In this way, these so-called catch muscles can keep the shells of the bivalve closed.

Two types of vertebrate smooth muscle

Vertebrate smooth muscles can be conveniently divided into single-unit and multiunit smooth muscles. In single-unit smooth muscle, all cells are connected to their neighbors by gap junctions, which allow the muscle cells to contract as a unit. As in the cardiac muscle, electric currents can be conducted across these gap junctions to induce contraction in the entire tissue. Also, like the cardiac muscle, single-unit smooth muscles have spontaneous activity. In contrast, multiunit smooth muscle cells have few gap junctions and operate individually when stimulated. They often have an extensive supply of nerves, and they can be activated or relaxed by nerves and hormones.

▶ *Single-unit smooth muscle is the most common form of smooth muscle. Nerves stimulate the single-unit muscle cells to contract by releasing neurotransmitters. The signal passes from one muscle cell to another through gap junctions, and the cells therefore contract as a unit. Multiunit smooth muscles are found in the uterus, the male reproductive tract, and in the eye. In multiunit muscle cells, each cell must be stimulated individually by neurotransmitters.*

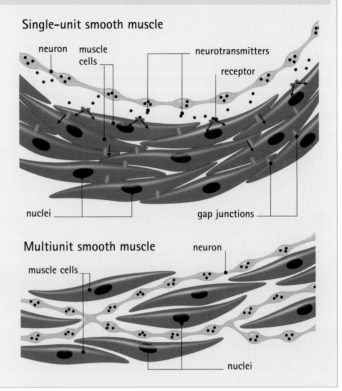

Single-unit smooth muscle

neuron muscle cells neurotransmitters receptor nuclei gap junctions

Multiunit smooth muscle

muscle cells neuron nuclei

and, when stimulation ceases, a slow removal of Ca^{2+} from the cytosol. Consequently, most smooth muscles contract and relax very slowly.

Smooth muscles do not have troponin. Instead, there are several other pathways that link an elevated Ca^{2+} concentration to myosin–actin cross-bridge formation. These pathways have been discovered mainly from chicken's gizzard. Vertebrates have at least six types of smooth muscles, including vascular and respiratory. Invertebrates have many other types. It is still not clear how pathways for contraction differ in different types of smooth muscles or animals. The current model is a combination of what is known from various smooth-muscle types and animals.

In smooth muscle, a rise in the cytoplasmic concentration of unbound Ca^{2+} is sensed by a protein called calmodulin. This protein removes an inhibitory protein from actin, exposing the myosin-binding sites on the actin filament. At the same time, calcium-bound calmodulin activates an enzyme that enables myosin to form cross bridges with actin and contract.

There is also evidence that binding of Ca^{2+} directly to myosin light-chain proteins induces a change in shape that allows binding of the myosin head to actin. Other mechanisms involve binding of Ca^{2+} to a calcium-dependent enzyme, called protein kinase. This enzyme is present in many cells, where it controls different processes, but in smooth muscle it can regulate contraction and relaxation according to the concentration of Ca^{2+} in the cytosol.

Unlike skeletal muscle, which contracts in response to nerve signals only, contraction or relaxation in smooth muscle is regulated by numerous factors, such as nerve signals and hormones. Furthermore, smooth muscle is often supplied with nerves from several different nerve fibers, each of which may have a distinct influence on the contraction state of the muscle.

CHRISTER HOGSTRAND

FURTHER READING AND RESEARCH
Muscolino, Joseph E. 2005. *The Muscular System Manual: The Skeletal Muscles of the Human Body, 2nd Edition.* CV Mosby: St Louis, MO.

Mushroom

KINGDOM: Fungi DIVISION: Basidiomycota
FAMILY: Agaricaceae

A mushroom is only a small part of the whole fungus to which it belongs. Mushrooms are the fruiting bodies of a fungal organism: they are equivalent to the apples on an apple tree. Most of the fungus is hidden inside wood or soil as microscopic threads called hyphae. These threads collectively form a mycelium. The term *mushroom* is generally used for any umbrella-shaped fungal fruiting body.

Fungi live anywhere there is a food source. Natural habitats include rotting wood, leaf litter, dung piles, ponds, tree bark, leaf surfaces, and even the insides of living plants and animals, especially insects. Fungi can also live in artificial habitats, such as shower curtains. Fungi span almost all scales of life, from microscopic spores and threads to organisms that are literally miles across.

Anatomy and taxonomy

Mushrooms form only a tiny part of the incredibly diverse fungal kingdom. The most familiar mushroom is the cultivated mushroom, *Agaricus bisporus*, but there are many other species with a similar structure. There are more than 70,000 species of fungi known to science, and probably more than 1 million species of fungi in total.

● **Fungi** Fungi were once considered to be simple plants, but it is now known that they have more similarities with animals. Fungi are placed in their own kingdom. They absorb nutrients from their immediate environment. Like plants, they are nonmobile for all or part of their life cycle. Some fungi, such as yeasts, consist of single cells, but more usually fungi consist of a network

▶ *Traditional fungal classifications are based mainly on the structure of the fruiting body and spore-producing tissues. Chemical and genetic analysis is adding to the understanding of how fungi are related, and changes to classification systems are being made all the time.*

of tubes called hyphae, bound by cell walls. The cell walls are stiffened with chitin—a polymer also found in the bodies of invertebrates such as insects. Plants have cellulose cell walls.

Some funguslike organisms are also found in the kingdom called protists, rather than in the fungi kingdom. These organisms, which include the slime molds and oomycota, have cell walls made of cellulose. Oomycota are water molds and downy mildews. Many, such as potato blight (*Phytophthora infestans*), cause plant diseases.

Fungi are often divided into higher and lower fungi. Higher fungi—the Ascomycota and Basidiomycota—are those that produce macroscopic fruiting bodies, whereas the lower fungi—the Zygomycota and Chitridiomycota—produce microscopic structures.

● **Chytrids** These lower fungi are predominantly aquatic organisms, composed of either single cells or a mycelium. The hyphae that make up the mycelium of chytrids have no septae (dividing walls), so the cell contents mingle together. Chytrids reproduce with mobile spores, each with a single, whiplike flagellum. They are the only fungi to have flagellae. Most chytrids are saprophytes (living on dead tissue), but

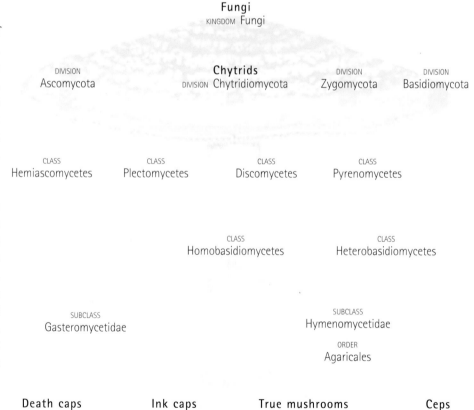

Fungi
KINGDOM Fungi

DIVISION Ascomycota

Chytrids DIVISION Chytridiomycota

DIVISION Zygomycota

DIVISION Basidiomycota

CLASS Hemiascomycetes

CLASS Plectomycetes

CLASS Discomycetes

CLASS Pyrenomycetes

CLASS Homobasidiomycetes

CLASS Heterobasidiomycetes

SUBCLASS Gasteromycetidae

SUBCLASS Hymenomycetidae

ORDER Agaricales

Death caps FAMILY Amanitaceae

Ink caps FAMILY Coprinaceae

True mushrooms FAMILY Agaricaceae

Ceps FAMILY Boletaceae

Cultivated mushroom GENUS AND SPECIES *Agaricus bisporus*

Common field mushroom GENUS AND SPECIES *Agaricus campestris*

some are pathogens (disease-causing organisms) of plants. One, *Batrachochytrium dendrobatitidis*, is a pathogen of frogs and may be responsible for their worldwide decline.

● **Zygomycota** Like most other fungi, the Zygomycota are mainly terrestrial. As with chytrids, their hyphae rarely have dividing walls. Sexual reproduction in Zygomycota species involves specialized branches of the hyphae, which join to form a zygospore. The order Mucorales includes the genus *Mucor*, which is a common bread mold.

● **Ascomycota** The Ascomycota are the largest division, or phylum, of fungi. The spores produced by sexual reproduction are formed within a sac called the ascus. These sacs are produced in fruiting bodies called ascocarps. The fungi are either single-celled yeasts or mycelial, with the hyphae divided into compartments by septae. The septae have a hole that allows organelles (functional parts) to move from one segment to another.

The classes of fungi within the Ascomycota include yeasts (Hemiascomycetes), powdery mildews (Plectomycetes), cup fungi (Discomycetes), and flask fungi (Pyrenomycetes). These different groups are identified usually by the arrangements of asci within the fruiting bodies.

● **Basidiomycota** This division of fungi includes mushrooms, bracket fungi, stinkhorns, puffballs, and jelly fungi. The hyphae in Basidiomycota fungi are divided by septae. Material moves between the mycelial compartments through the septae via a complicated structure called a dolipore septum. The spores in Basidiomycota fungi are produced on a structure called a basidium.

● **Heterobasidiomycetes** This class of fungi includes the rust fungi (Uredinales), smut fungi (Ustilaginales), and jelly fungi (Tulasnellales). Smut and rust fungi, such as *Puccinia*

▲ *Many colorful mushrooms, such as this fly agaric, are called toadstools. That name has no taxonomic meaning, but in general, toadstools are poisonous mushrooms and some may kill if eaten.*

and *Ustilago*, are parasitic on plants and have no fruiting body. The jelly fungi, for example *Tremella* and *Auricularia*, have fruiting bodies with a gelatinous texture.

● **Homobasidiomycetes** This class of fungi usually forms fruiting bodies, and the class contains the familiar mushrooms and other large fruiting bodies in a variety of shapes. In Aphyllophorales, the spores are produced on spines, tubes, wrinkles, or a smooth surface. This order of fungi within Homobasidiomycetes includes bracket fungi, club fungi, and crust fungi. Puffballs, earth stars, and bird's nest fungi are also part of the Homobasidiomycetes class. In stinkhorns, another Homobasidiomycetes family, the spores are in a gooey mass on top of a stalk.

● **Agaricales** The Agaricales order contains the fungi with fleshy fruiting bodies of a classic mushroom shape, usually with a cap, stem, and gills (or tubes in boletes fungi). In ink caps, the whole cap dissolves into a liquid spore mass. The death-cap genus includes many poisonous species, including death cap (*Amanita phalloides*), destroying angel (*Amanita virosa*), and fly agaric (*Amanita muscaria*).

The Agaricaceae family includes the parasol mushrooms and true mushrooms. The familiar cultivated mushroom, *Agaricus bisporus*, is a two-spored version of the common field mushroom, *Agaricus campestris*.

FEATURED SYSTEMS

EXTERNAL ANATOMY A mushroom is the fruiting body of a much larger hidden structure (a mycelium), which is composed of tiny threads (hyphae). The mushroom itself has a stem and a cap. Spores drop from the gills on the underside of the cap. *See pages 772–775.*

INTERNAL ANATOMY The threadlike hyphal tubes are not divided into separate cells. Instead, many nuclei and organelles share cytoplasm (the cell contents). There are occasional dividing walls called septae, which have pores. *See page 776.*

REPRODUCTION Fungi reproduce sexually and asexually using spores. The many types of spores and their supporting structures are used to classify fungi into different groups. *See pages 777–779.*

External anatomy

COMPARE a fungal hypha with an *AMOEBA*. Both feed on complex organic (living or once-living) materials. The amoeba engulfs food particles and digests them internally, whereas the fungus excretes enzymes and absorbs the digested products.

CONNECTIONS

Fungi can be huge. One *Armillarea bulbosa* fungus in Michigan extends through a forest over 37 acres (15 hectares) and is estimated to weigh 110 tons (100 metric tons)—almost as much as a blue whale. This single organism grew from a spore that germinated more than 1,000 years ago.

Mushrooms are only a small part of the total organism. The rest exists as thin tubes that spread through the fungus's habitat. When a spore lands on a suitable surface, such as soil, wood, or leaf litter, it germinates. Hyphae—microscopic tubes, usually between 1 and 15 micrometers (μm, millionths of a meter) across, filled with cytoplasm (cell contents)—grow out from the spore. The hyphae keep growing and branching to form a mycelium, which has the potential to grow forever. Sections of the mycelium that become separated from the rest will also survive and grow independently.

▶ Fly agaric
The fly agaric is a common mushroom found throughout the Northern Hemisphere in damp forests and meadows. This poisonous mushroom is so named because it was once used to make fly poison.

▼ GILLS FROM BELOW
The gills radiate from the stem underneath the cap. They are covered in a layer of spore-producing cells called basidia. The spores are released into the air and blown away by the wind.

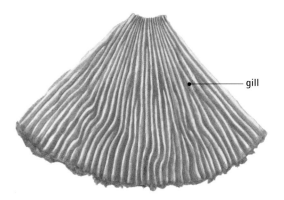

gill

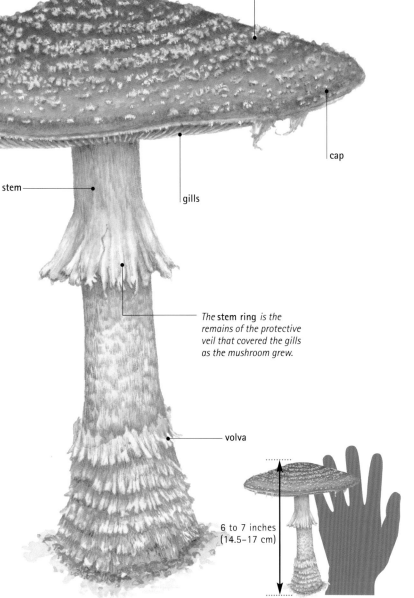

cap scale

cap

stem

gills

The **stem ring** *is the remains of the protective veil that covered the gills as the mushroom grew.*

volva

6 to 7 inches
(14.5–17 cm)

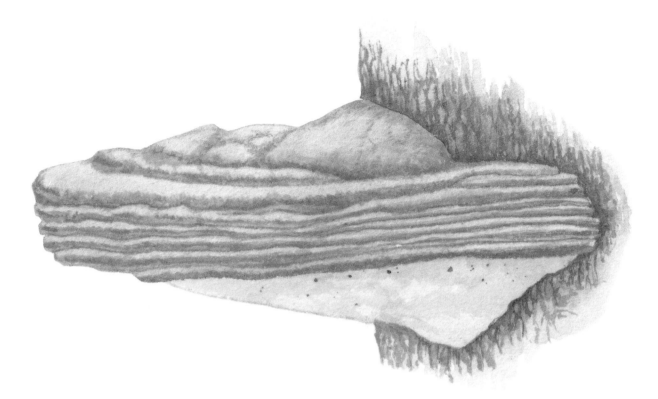

The mycelium is the feeding portion of the mushroom. Its function is to absorb food and use it to power the mushroom's life processes or turn it into more fungal tissue. The thin branching growth form is good for exploring large areas using the mininum amount of body tissue.

Unlike plants, fungi do not photosynthesize. (Photosynthesis is a process in which energy from sunlight is harnessed to make food from carbon dioxide and water.) So, like animals, fungi obtain food by breaking down complex substances made by other organisms.

▲ **Bracket fungus**
Bracket fungi, such as this Ganoderma *species, form shelflike growths on trees and wooden structures.*

Fungi in partnerships

Fungi form partnerships with many other organisms. Lichens are a joint organism—algal cells nestled in a mesh of fungal mycelia. Fungi also have close mutually beneficial associations with plants and animals.

Mycorrhizal fungi grow around and even inside the plant roots, feeding on sugars that the plant produces from photosynthesis. In return, the mycelia act as root extensions, spreading far out into the soil and collecting water, phosphorus, and other minerals that the plant could not reach on its own.

Leaf-cutting ants cultivate Lepiotaceae fungi for food. The ants feed "gardens" of fungus with chewed-up leaves and then eat the knobbly structures, called bromatia, that develop on the mycelium.

▼ YOUNG BRANCHING MYCELIUM
The mycelium is the network of hyphal tubes that develop from a single spore. As the mycelium grows, the hyphae elongate and repeatedly divide into branches.

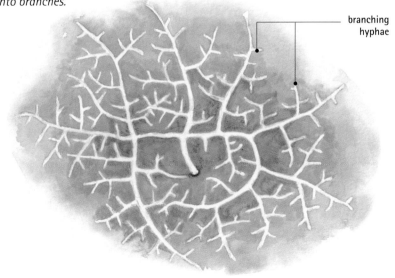

branching hyphae

▶ **Cup fungus**
Some fungi in the division Ascomycota are called cup fungi because they grow a cup-shape fruiting body. Spores are produced in sacs, called asci, which grow on the upper surface. The rare edible fungi, truffles and morels, are cup fungi, although their fruiting bodies are not typical.

There are three major feeding types of fungi. Saprotrophs feed on dead material. These fungi decompose dead bodies and play a vital role in ecosystems. Necrotrophs are parasites that directly kill the tissue of a host organism, then feed on the remains. The hosts can be plants, other fungi, invertebrates, and even mammals. Biotrophs feed on a living host. This relationship may be parasitic, where the host suffers; or mutualistic, where the host also benefits.

Mushroom structure

When two hyphae from compatible mycelia meet, they fuse. Budlike structures called primordia form, and the fruiting body, such as a mushroom, develops from these primordia.

A mushroom is made up of a cap, stem, and gills. The cap shelters and protects the spore-producing gills on the underside. The stem lifts the cap and gills above ground level, so that air can circulate around the gills to disperse the

CLOSE-UP

Nematode catchers

Some fungi prey on nematodes (roundworms) living in soil and dung. The fungi use several mechanisms for catching the worms, ranging from glue to nooses that strangle their victims. In *Arthrobotrys*, the entire mycelium is a sticky, tangled mesh. When a nematode gets stuck, a thin penetrating hypha grows into the living animal. A mass of absorptive hyphae then grow throughout the nematode and digest it from the inside. In *Dactylaria brochopaga*, hyphal rings form an active trap. They are touch-sensitive, and as soon as a nematode enters a loop the ring contracts like a noose. The reaction is fast—within a tenth of a second.

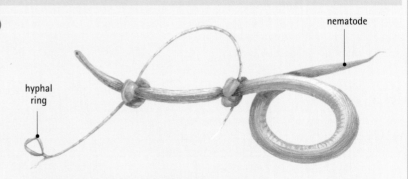

▲ **FUNGUS CATCHING A NEMATODE WORM**
Members of the Dactylaria genus catch tiny nematode worms in nooselike hyphal rings. When a worm slithers into a ring the noose tightens. The fungus then grows into the animal's body and digests it.

tiny spores. The gills are thin plates that hang vertically from the underside of the cap. They radiate from the center, and in some species they are branched. The gill surface is covered by a hymenium. This is a single layer of club-shaped, spore-producing cells called basidia. Each basidium typically forms four spores.

Some mushrooms have one or more veil layers that cover and protect the developing mushroom. The veil is a thin bag that breaks open as the mushroom expands, leaving traces on the mature fruiting body. These traces can take the form of a ring round the stem, loose scales on the cap surface, or a volva (cuplike structure) at the base. Cultivated mushrooms have a stem ring; the fly agaric mushroom also has the volva and cap scales—the white spots on the red cap.

The umbrella-shaped fruiting body of the mushroom-producing fungi comes in a variety of sizes, textures, and colors. Certain mushrooms in tropical forests have caps less than 0.03 inches (1 mm) across, whereas an East African mushroom called *Termitomyces titanicus* grows to 3.2 feet (1 m) across.

Fruiting body shapes

Fungal fruiting bodies also come in other shapes. For example, bracket fungi, which grow on tree trunks or stumps, produce spores on the walls of closely packed vertical tubes, each opening to the underside of the fruiting body. Some species produce so many spores that the stump below is covered in rust-colored dust.

In Ascomycota fungi, spores are usually produced simply on the fruiting body's surface, rather than on protected gills. In cup fungi, the fruiting body is a cup-shaped structure with asci (spore sacs) on the inner surface. Truffles are irregular, lumpy fruiting bodies of the *Tuber* genus that form underground. Their inner surface is covered with asci.

▲ *The distinctive, musty flavor of truffles makes them a highly prized edible fungus. It is difficult to cultivate truffles, and most are harvested from the wild. Truffle hunters use hogs or dogs to sniff them out.*

Internal anatomy

Like plants, animals, and protists, fungi are eukaryotes: their cells are filled with structures bound by membranes that carry out a range of different jobs. For example, a yeast has internal structures similar to any other typical eukaryotic cell, with all the standard organelles, including the nucleus, ribosomes, mitochondria, and endoplasmic reticulum, contained by a cell membrane.

Like other fungi, mushroom-bearing species are composed of a mass of filamentous hyphae, which make up the mycelium. In the body of a fungus—unlike the bodies of plants and animals—the hyphae that make up the body are not divided into individual cells. Instead, the hyphae are filled with a continuous cytoplasm that contains many nuclei and other organelles.

The hyphal wall is strengthened with fibers of chitin. This is the same molecule that strengthens the bodies of insects, crabs, and other invertebrates. In the higher fungi, the hyphae are divided by walls called septae. Septae strengthen the hyphae. Cellular materials can move through the septae. However, the septae can also seal off areas of the mycelium to restrict damage and to allow different parts to develop in different ways, such as growing fruiting bodies—for example, a mushroom.

The "magical" growth of a mushroom

Plants and animals grow by cell division and multiplication. This process is relatively slow and cannot match that of some fungi, in which the mushrooms appear overnight. These rapid growth rates have given rise to many myths about mushrooms.

Mushroom fruiting bodies grow from a buttonlike structure called a primordium. Inside the primordium, cells have divided to produce thousands of small hyphae, all arranged in the form of a miniature mushroom. When conditions are right, these hyphae take up water and inflate enormously, so the fruiting body expands rapidly.

Septum structure

The septum is a barrier with a pore about 2 μm across. Ascomycetes have just a simple pore, which is large enough to allow the organelles to move through it.

Basidiomycetes have a more complex pore, the dolipore septum, in which each pore forms a doughnut-shape ring. Surrounding the open ends of the tube is an endoplasmic-reticulum complex called a parenthosome. This is so named because it cups the pore at each end like a pair of brackets or parentheses.

▼ **RHIZOMORPH**
A rhizomorph is a bundle of hyphae, commonly found attached to the base of fruiting bodies. The vessel hyphae are often empty, allowing the easy flow of nutrients. The dark brown area may contain very narrow hyphae. The chitin wall thins near the growing tip.

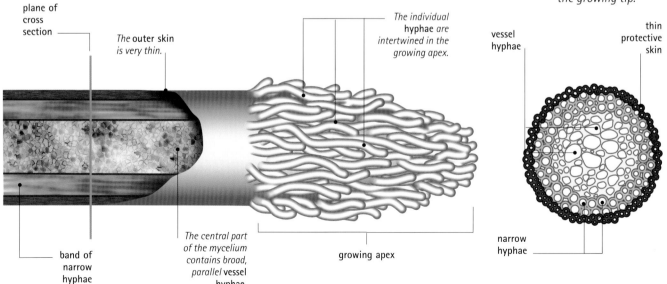

plane of cross section

The outer skin *is very thin.*

The individual hyphae *are intertwined in the growing apex.*

vessel hyphae

thin protective skin

band of narrow hyphae

The central part of the mycelium contains broad, parallel vessel hyphae.

growing apex

narrow hyphae

Reproduction

Fungi reproduce by using spores. Spores are small, dispersible cells that develop according to a controlled process, known as germination. Spores can be generated by sexual or asexual processes, and most fungi undergo both sexual and asexual reproduction at different stages in their life cycle.

In sexual reproduction the genetic material from two parents fuses into a single genetically unique cell. This requires parent cells to divide by meiosis, so the cell nucleus, with the genes it contains, is split into two halves. Half sets of genes from two parents then merge to make a spore containing a unique full set.

Asexual reproduction does not involve changes to the genetic material in order to make new individuals. Instead, it is used to disperse an existing genome—the set of genetic instructions for an already existing organism. This dispersion can happen over time, as well as distance: certain thick-walled spores can survive in a dormant state for many years until environmental conditions change.

Asexual reproduction

There are two main types of asexual spores. Sporangiospores are produced within a saclike swelling of the hyphae. Conidia are held outside the hyphae and occur in a variety of forms. The *Penicillium* species, for example, have branching clusters of conidia at the ends of hyphae.

Yeasts reproduce asexually, but by budding and fission rather than by spore production. In budding, smaller offspring cells pinch off from the parent, whereas in fission the parent cell splits in half. Both processes depend on mitosis. The genes in the parent cell duplicate and divide, so offspring cells have the same genes as the parent.

Sexual reproduction

There are three essential phases in fungal sexual reproduction. Plasmogamy is the fusion of separate hyphae—the equivalent of the fusion of cells in other organisms. In karyogamy, two haploid nuclei—a nucleus with only a single set of chromosomes—fuse

▶ LIFE CYCLE

Methods of sexual reproduction vary in different types of fungi. In those belonging to the Basidiomycota, hyphae from two different spores fuse to form a hypha with two nuclei. This grows into a fruiting body, or mushroom. In the mushroom's gills, some of the paired nuclei fuse and form special reproductive structures called basidia (singular, basidium). These cells with fused nuclei are called diploid (cells with a full complement of DNA). Soon afterward, a process called meiosis occurs in which the nuclei divide, producing four new nuclei, each with half the full complement of DNA. These nuclei migrate to the surface of the basidium and form four spores. These spores then fall or are blown from the basidium; and if they settle in a suitable environment, they will germinate and grow into hyphae.

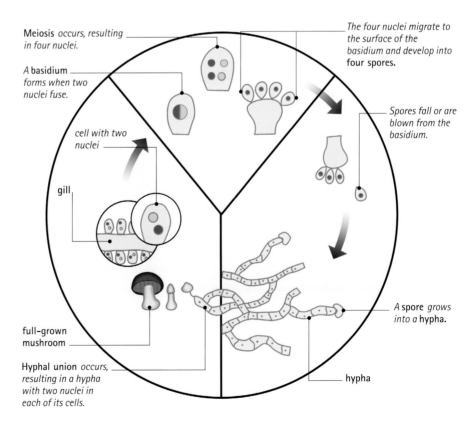

Meiosis *occurs, resulting in four nuclei.*

A basidium *forms when two nuclei fuse.*

cell with two nuclei

gill

full-grown mushroom

Hyphal union *occurs, resulting in a hypha with two nuclei in each of its cells.*

The four nuclei migrate to the surface of the basidium *and develop into* four spores.

Spores *fall or are blown from the* basidium.

A spore *grows into a* hypha.

hypha

777

to form a diploid nucleus, one with a full complement of paired chromosomes. Finally, meiosis separates paired chromosomes, so that the offspring cells have only a single set of chromosomes and return to the haploid state.

Some types of fungal mycelium are capable of completing a sexual cycle without fusing with another individual, a process equivalent to self-fertilization in plants. In others, mycelia from two different individuals have to be involved to complete the cycle.

In some cases, the two mycelia have different structures, and can be thought of as male and

▼ *The wolf-fart puffball,* Lycoperdon pyriforme, *releases spores when it is hit by raindrops or knocked by passing animals. Each spore is less than 5 μm wide.*

female. In others, sex organs cannot be identified, so the individuals are assigned to different "mating types."

Shaped for dispersal

The huge variety of fruiting-body shapes, from mushrooms to brackets, crusts, antlers, and cups, reflects the many ways fungi disperse their spores, including by wind, water, and insects.

Among these methods of dispersal, wind dispersal is most common. Mushrooms and other fungi with gills and tubes tend to be wind-dispersed. Wind-dispersed spores are small, light, and dry, often with surface projections (bumps and spines). Billions of spores can be produced at a time. For example, the fruiting body of a giant puffball (about the size a soccer ball) can produce 7 trillion spores.

The stalk of mushrooms and other fruiting bodies lifts the spores above ground level. The hyemenium is on the gills, hanging underneath the cap, so the spores can drop into circulating air currents. Overhanging bracket fungi use the same principle.

Some fungi, however, use the energy of raindrops to propel spores. Puffballs have a strong elastic skin that recoils when a raindrop hits it, puffing out its contents of spores. Bird's nest fungi are named for their nestlike fruiting bodies, which contains egg-shaped spore packets, called peridioles. The cupped shape of the "nest" directs any droplets falling into it so that a peridiole is splashed out of the cup.

Other fungi use an explosive method for dispersing their spores. *Pilobolus* is a pin mold that lives on dung heaps. The spores are in a

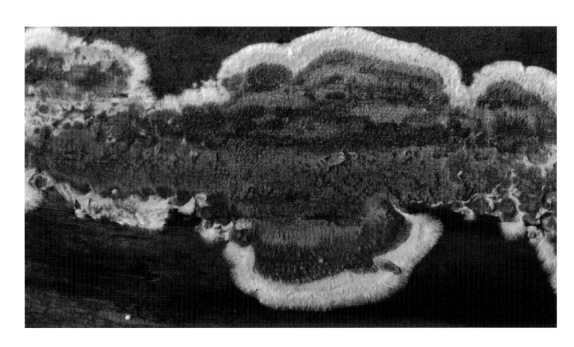

◀ *Crust fungi such as this* Coniophora *grow as a layer on the surface of wood. The white fur at the edges is mycelia.*

▼ *The mushroom fruiting body grows from a mycelium. A network of hyphae extends through the fruiting body. Spore-producing structures are arranged on the mushroom's gills.*

sporangium. This takes the form of a black blob that sits on a clear vesicle, on top of a long stalk. Water builds up inside the vesicle so that, at maturity, it resembles a tightly filled balloon. Any slight trigger causes it to burst and propel the entire sporangium at an initial speed of about 30 miles per hour (50 km/h), up to 6.4 feet (2 m) away.

The stinkhorn, *Phallus impudicus*, also uses a notable method of dispersal. It attracts flies with its unpleasant odor. The fruiting body has an upright, hollow stem bearing a cap on which a slimy spore mass, or gleba, is produced. Within a few hours of the fruiting body's emergence, the dark, gooey spore mass has been completely removed by the visiting flies.

ERICA BOWER

GENETICS

Mating types

In Ascomycetes, mating type is controlled by a single gene. There are two different alleles (varieties) of the gene. For two mycelia to be compatible, they need to have different alleles. In Basidiomycetes, mating type is controlled by one or two genes, each of which can have many different alleles. There can be hundreds of variations. This is equivalent to having hundreds of sexes, rather than just male and female.

hyphae

cap *of fruiting body*

hyphae

spore-producing structures *such as* basidia

A mass of hyphae form a mycelium.

FURTHER READING AND RESEARCH

Hawksworth, D. L., P. M. Kirk, B. C. Sutton, and D. N. Pegler. 1995. *Ainsworth and Bisby's Dictionary of the Fungi*. CAB International: Wallingford, UK.

779

Nervous system

All organisms receive information, process it, and then produce an appropriate response. In complex animals, these functions are, for the most part, performed by two interconnected systems: the nervous system, and the endocrine system, which produces hormones.

The nervous system consists of often huge networks of nerve cells that perform three interconnecting functions. First, the nervous system allows animals to sense what is happening in their environment. Their environment is both internal (within the body, including such things as hormonal levels or the amount of stretch in a muscle) and external (outside the body—for example, as monitored by vision, hearing, and touch). Second, the nervous system processes this sensory information and compares the information from different senses. The processing can be a relatively simple reflex or extremely complex, as in human speech. Third, the nervous system enables animals to do things, primarily by controlling muscles and glands. The three functions can be accomplished amazingly quickly, within a few milliseconds (thousandths of a second). This speed of information transmission is achieved by electrical and chemical signals within and between nerve cells.

Processing the vast amount of sensory information and producing an appropriate response require nervous systems to be remarkably complicated. This complexity can be seen at all levels. For example, how individual molecules of the nervous system operate is a scientific field in itself (called biophysics). Most nervous systems have huge numbers of nerve cells and even more connections between them. Such complexity is necessary to produce the subtle range of behaviors most animals display.

Different types of nervous systems

The first nervous systems to evolve were probably networks of identical nerve cells that connected different parts of the body and allowed a general response to limited sensory input. Later, nerve cells specialized and formed physical groups for specific functions. Such groups are called ganglia (singular, ganglion). For example, in segmented animals, such as earthworms, there are ganglia in each segment. In gastropod mollusks, such as slugs and snails, the nervous system includes a buccal (mouth) ganglion, which helps control feeding; and a pedal ganglion, which is involved in locomotion. Sponges are the only multicellular animals that do not possess a nervous system.

Most animals tend to move mostly in one direction. The front end (the head of the animal) tends to encounter

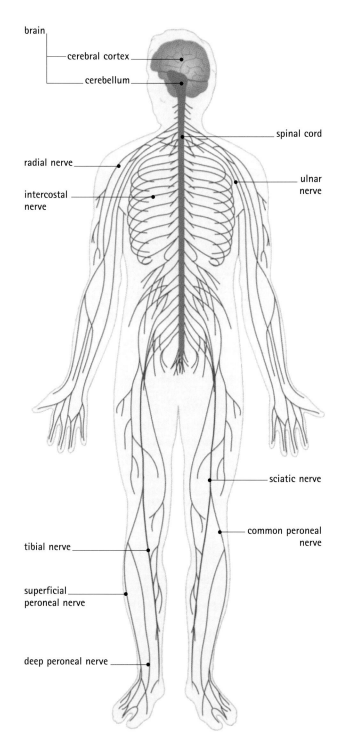

▲ NERVOUS SYSTEM
Human
Typical of vertebrates, humans have a nervous system made up of a central nervous system (the brain and spinal cord, shown in orange) and the peripheral nervous system (a network of nerve fibers branching to every part of the body, shown in blue).

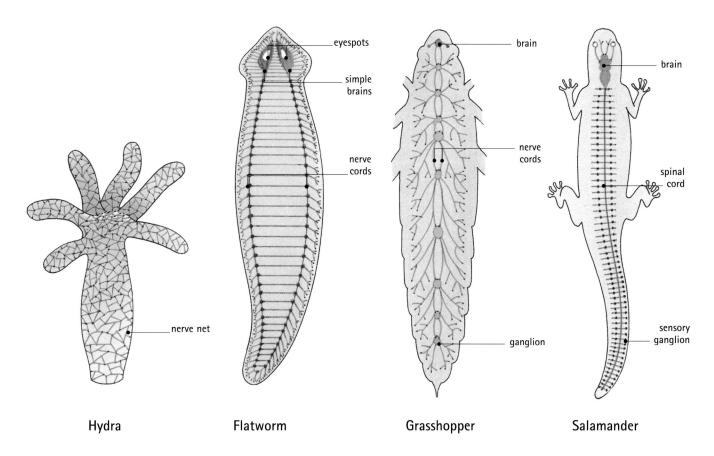

Hydra   Flatworm   Grasshopper   Salamander

▲ *From left to right, these animals have a progressively more complex nervous system, from a hydra with a loose nerve net to a salamander with a brain and spinal cord.*

FEATURED SYSTEMS

HOW THE NERVOUS SYSTEM WORKS The nervous system is made up of nerve cells (neurons). There are various anatomical types of neurons to match their specialized functions. Neurons deal with information—receiving it, transmitting it, and relaying it to other neurons, muscles, or glands—by using electrical and chemical signals. *See pages 782–785.*

SENSORY AND MOTOR SYSTEMS Sensory systems allow animals to see, hear, feel, taste, and smell. There are other senses as well: some animals can detect electrical or magnetic fields. Motor systems allow animals to make appropriate responses based on sensory input. *See pages 786–791.*

INVERTEBRATE NERVOUS SYSTEMS An extraordinary diversity of nervous systems—as diverse as the group itself—occurs in invertebrates. *See pages 792–793.*

VERTEBRATE NERVOUS SYSTEMS Vertebrates have a central nervous system—consisting of a brain and spinal cord—and a peripheral nervous system that connects to sensory organs and muscles. *See pages 794–799.*

stimuli first, and most sensory structures are of greatest use there. All this sensory input is therefore dealt with by the frontmost ganglion, so it has evolved to be larger and have more nerve cells than the others. This large front ganglion is often called the brain. Brains of one type or another are found in both invertebrates and vertebrates. So, in most animals there is some form of brain at the front end and a nerve cord or cords (with ganglia) running the length of the body that collect sensory information and distribute motor commands locally. Within this basic form, considerable variations occur across the animal kingdom. For example, in insects two parallel nerve cords run along the belly, whereas in vertebrates a single cord runs along the back.

The nervous system of vertebrates may be divided into two main parts—the central nervous system (CNS), which consists of the brain and spinal cord; and the peripheral nervous system, which comprises the nerves that connect with the brain and spinal cord.

All nervous systems are complicated, and the largest in terms of numbers of nerve cells (about 100 billion in the human brain) and the hardest to understand is the human nervous system. Unraveling the workings of the human brain, which is much more complicated than even the most sophisticated computer, remains one of the greatest challenges to scientists.

How the nervous system works

COMPARE the nervous system with the **ENDOCRINE AND EXOCRINE SYSTEMS**. The nervous and endocrine systems work closely together and interact strongly. Nerve cell activity can control hormone release, and in turn hormones can affect the nervous system. The endocrine system usually operates more slowly (from seconds to years) than the nervous system, which generally takes fractions of a second to work.

The word *neuron* is another name for a nerve cell. Neurons have the same basic anatomy as other body cells. They have a nucleus containing genetic material (DNA), have cytoplasm with organelles, and are surrounded by a cell membrane. Some features are common to most neurons. The nucleus is contained within a cell body, or soma. There are a number of branching structures called dendrites, which receive information from other neurons; and one or more long filament-like extensions called axons, which send information. A nerve is simply a bundle of axons that serve the same part of the body.

Neurons perform many tasks within the nervous system, so their anatomy is amazingly varied to match their specialized functions.

Because they have to send information over distances, some neurons are the longest cells in the body and can be more than 3 feet (1 m) long in humans. Neurons can be broadly divided into three classes. Sensory neurons receive and then transmit information from sense organs and cells. Motor neurons send the output of the nervous system to muscles and glands. Interneurons pass information between neurons and are the most common type of neuron in the nervous system.

For the nervous system to work, neurons need to receive, process, and pass on information. This capability requires a means of sending signals over long distances within individual neurons, along their axon. It also needs a way for signals to pass between neurons.

▶ **RESPONDING TO STIMULI**

In vertebrates, stimuli detected both outside and inside the body produce information that is transmitted by sensory neurons to the central nervous system. There, the information is interpreted, bringing about an appropriate response, by motor, or efferent, neurons.

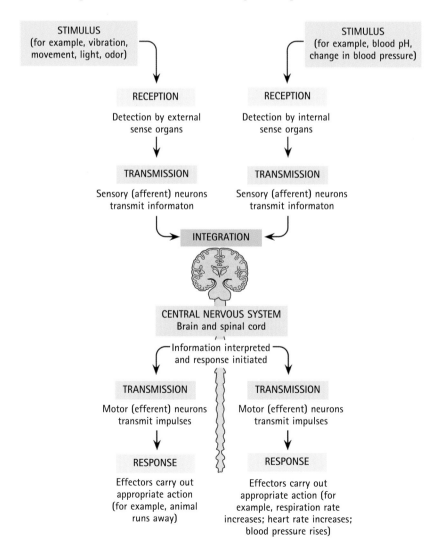

STIMULUS (for example, vibration, movement, light, odor)

STIMULUS (for example, blood pH, change in blood pressure)

RECEPTION — Detection by external sense organs

RECEPTION — Detection by internal sense organs

TRANSMISSION — Sensory (afferent) neurons transmit informaton

TRANSMISSION — Sensory (afferent) neurons transmit informaton

INTEGRATION

CENTRAL NERVOUS SYSTEM Brain and spinal cord

Information interpreted and response initiated

TRANSMISSION — Motor (efferent) neurons transmit impulses

TRANSMISSION — Motor (efferent) neurons transmit impulses

RESPONSE — Effectors carry out appropriate action (for example, animal runs away)

RESPONSE — Effectors carry out appropriate action (for example, respiration rate increases; heart rate increases; blood pressure rises)

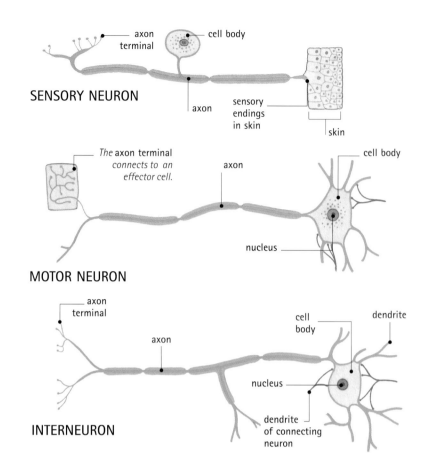

SENSORY NEURON

axon terminal — cell body — axon — sensory endings in skin — skin

MOTOR NEURON

The axon terminal connects to an effector cell. — axon — cell body — nucleus

INTERNEURON

axon terminal — axon — cell body — dendrite — nucleus — dendrite of connecting neuron

How signals are transmitted

The concentrations of electrically charged ions vary on either side of a neuron's membrane, creating a voltage difference across the cell membrane. At rest, this voltage difference is called the resting membrane potential and is about −70 millivolts (mV). In other words, the inside of the neuron is 70 mV more negative than the outside, and the membrane is said to be "polarized." This difference results because there is a higher concentration of positively charged sodium ions outside the cell, and inside the cell a higher concentration of negatively charged chloride ions and a lower concentration of positively charged potassium ions. Protein molecules in the cell membrane maintain this difference by actively pumping ions from one side of the membrane to the other.

Electric signals are transmitted in neurons when there is a change in the membrane potential—that is, when there is a movement of ions across the cell membrane. The most common form of transmission is called an action potential, or nerve impulse. If a stimulus causes the membrane potential to become less negative

▲ THREE TYPES OF NEURONS

Sensory neurons detect and transmit sensory information. Motor neurons carry impulses to effector cells such as muscles. Interneurons pass information between neurons.

Squid giant axons

Many early studies on neuronal properties used neurons that mediate the rapid escape response in squid. These neurons have axons that are very large—up to 0.04 inch (1 mm) in diameter. Because of their large size, isolated squid axons can have metal electrodes inserted into them to measure the membrane potential and how it responds to different stimuli. The contents of the axon can also be squeezed out, like toothpaste from a tube. This allows the concentrations of various ions within the axon to be measured and compared with concentrations outside the axon.

(depolarized), an action potential is triggered. That happens, however, only if the stimulus is strong enough to breach a certain threshold.

At the start of an action potential, ion channels in part of the membrane allow an inward rush of sodium ions. The local membrane potential momentarily becomes positive, passing 0 mV. The sodium channels then begin to close, and potassium ions followed by sodium ions are pumped out of the cell. Then, there is a rapid return to the resting membrane potential at the site of depolarization. The whole process of generating an action potential takes only 2 to 5 milliseconds.

When a region of cell membrane triggers an action potential, it creates a stimulus that depolarizes a neighboring region of membrane. In this way, the action potential is propagated,

Venoms that attack the nervous system

Venoms often contain specific molecules that act against the nervous system. Some of these molecules (such as tetrodotoxin from pufferfish) stop neurons from producing action potentials. Others (like bungarotoxin from snakes called kraits) block the receptor molecules at synapses (junctions between neurons) and stop communication between neurons. For victims of the venom, this can have painful or even lethal consequences. However, when isolated, these molecules can be used in experiments to study the nervous system. By using venom molecules to block a specific activity, scientists can investigate its normal function.

Electric plants

Neurons are not the only cells that are electrically excitable. Muscle cells also have a resting potential that responds to electrical input from neurons. In addition, some plant cells have resting potentials and can produce the equivalent of action potentials—for example, the cells involved in the fast movement with which the insectivorous Venus flytrap captures its prey.

or travels, along the neuron—usually starting at the cell body and then along the axon—at up to 200 miles (320 km) an hour. Action potentials are all-or-nothing, so they cannot vary in size. Information is therefore transmitted by varying the frequency and timing of action potentials.

How one neuron connects with another

For the nervous system to be efficient, signals need to pass between neurons. Information is passed between neurons at junctions called synapses. At most synapses, the axon of one neuron makes one or more contacts with the dendrites or cell body of another. There may be more than 1,000 synapses on the cell body and

▶ **NERVE IMPULSE**
At rest, there is a higher concentration of negatively charged ions inside a neuron than outside, creating a voltage potential (difference) across the membrane (1). A stimulus above a certain threshold level depolarizes a region of cell membrane and triggers an action potential, in which there is a local influx of sodium ions into the cell (2). Sodium and potassium ions are then pumped out of the axon at the site of depolarization, and the membrane returns to the resting potential; however, the neighboring region of membrane is then sufficiently depolarized to trigger another influx of sodium ions; and in this way the action potential is transmitted along the axon (3).

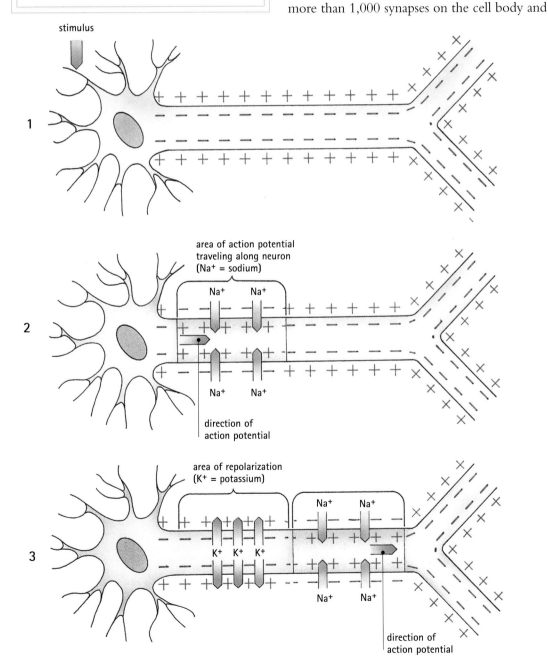

Holding it all together

Neurons cannot work on their own. They are assisted by a range of cells called glial (Greek for "glue") cells. Glial cells have various roles, from physical support to electrical insulation. Although (unlike neurons) they cannot signal rapidly over long distances, glial cells are vital for the nervous system to work. In fact, glial cells are 10 to 50 times more numerous than neurons. Some glial cells even provide growth factors, which are chemicals essential for some neurons to survive. One important type of glial cell in vertebrates is the Schwann cell. Schwann cells make a form of protective sheath around axons. However, there are small uninsulated gaps at regular intervals. There, the action potential jumps from one gap to the next, allowing it to travel much faster.

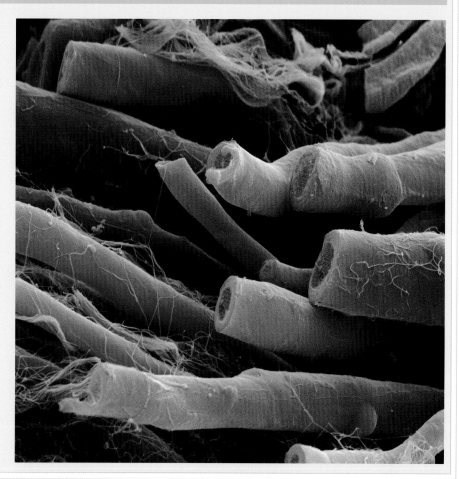

▶ *This scanning electron micrograph shows axons surrounded by a protective sheath (purple) that is formed of concentric layers of Schwann cells.*

dendrites of just one human motor neuron. The most common type of synapse is the chemical synapse. The neuron making contact has a number of vesicles, which are internal spheres of membrane that contain molecules of a chemical messenger called a neurotransmitter. A neurotransmitter is a specific molecule that transfers a signal between neurons. Two examples of neurotransmitters are dopamine and acetylcholine. When an action potential arrives at a synapse, the vesicles fuse with the cell membrane and release the neuro-transmitter into the synaptic cleft, the tiny gap between a neuron and its target cell. The neurotransmitter molecules diffuse rapidly across the narrow cleft and then bind to receptor molecules on the cell membrane of the receiving neuron. Receptor molecules are specific to individual neurotransmitters and can

work in a variety of ways. Some receptors open or close channels in the membrane, allowing certain electrically charged ions to enter or leave the neuron. Others work through chemical signaling systems within the neuron. The final result is a change in the membrane potential, either positive or negative, of the receiving neuron. Individual neurons usually receive information from numerous synapses. If the sum of the inputs depolarizes the neuron sufficiently, it will trigger one or more action potentials, passing on an electrical signal.

At an electrical synapse the two neurons make a direct electrical contact. Transmission at an electrical synapse is fast, but there is not much scope for modifying the signal, as there is at a chemical synapse. Modifying signals at chemical synapses is necessary for processes like learning and memory.

Sensory and motor systems

Sensory systems allow animals to detect conditions inside and outside the body, in addition to any changes in those conditions. To detect such information, energy from the sensory stimulus, such as photons of light or vibrations in the air, or even more direct stimuli such as touch, must be converted into a signal the nervous system can distribute and process. Sensory information is first detected in cells called sensory receptors. Receptors are usually modified neurons or epithelial cells on the outer surface of the body and the walls of internal cavities. Receptor cells transform the energy of the sensory stimulus into an electrical signal. This electrical signal is then transmitted to the CNS as a series of impulses called action potentials, initiated either in the receptor or in neurons connected to it.

Receptor cells will detect sensory signals only within a certain range. For example, humans can detect light only between the wavelengths of 380 and 760 nanometers (0.00000038 and 0.00000076 m) and sound between frequencies of 20 and 20,000 hertz (Hz, air vibrations per second). Different animals can have different sensory ranges. Bats can hear sound frequencies as high as 120,000 Hz (ultrasound), whereas elephants can hear frequencies as low as 1 Hz (infrasound). Often, different receptor types are used to detect different ranges of a sensory stimulus. In the human eye there are three types of receptors responsible for color vision, which are sensitive to red, green, or blue wavelengths of light. The brain compares the information from these different receptors to achieve a full perception of color. Receptors can also register the strength (intensity) of a sensory stimulus. A more intense stimulus results in more action potentials transmitted close together, and in more receptors becoming active.

Sensory processing
Although the sensory stimulus may vary (touch, pain, light, sound), the final input received by the nervous system is always a series of action potentials. This sensory signal is then converted into a perception of a sense in the brain. So you do not really "see with your eyes," but in fact form a perception of vision in your brain. Transforming a series of action potentials into a perception of the environment starts at the receptor cell itself. When a change in the environment occurs, a strong neuronal signal is usually generated (many, frequent action potentials). If the environment then maintains this new configuration, the signal will reduce over time. The number and frequency of action potentials will be reduced. This is a process known as sensory adaptation. Note that this is quite separate from the meaning of adaptation in an evolutionary sense. One example of sensory adaptation is putting on a wristwatch: when you first put a watch on your wrist you can feel it, but you become less aware of it over time. The advantage of adaptation is

▼ *This false-color scanning electron micrograph shows photoreceptors in the human retina. The tall brown structures are rods (responsible for vision in dim light), and the less common, short orange structures are cones (responsible for color vision). Magnified 3,300 times.*

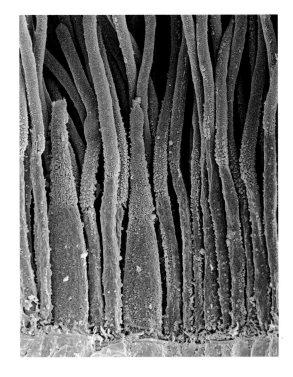

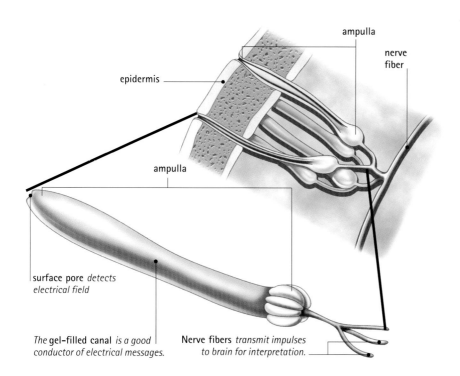

epidermis

ampulla

nerve fiber

ampulla

surface pore *detects electrical field*

The **gel-filled canal** *is a good conductor of electrical messages.*

Nerve fibers *transmit impulses to brain for interpretation.*

◀ **AMPULLAE OF LORENZINI**
Hammerhead shark
Sharks and rays have electroreceptors called ampullae of Lorenzini embedded in their head. These receptors allow them to detect the electrical field of other aquatic animals, including prey.

that it provides the organism with information about changes in the environment without overloading the nervous system. Much more sophisticated sensory processing occurs in the CNS. This allows us, for example, to interpret different wavelengths of light as different colors. The higher processing and integration of sensory information is a complex issue and one that is only just starting to be understood.

Sensory receptor cells can occur singly or in large groups. Receptors are usually part of a larger structure called a sensory organ. Sensory organs range from simple individual spines on a cockroach leg, which detect touch, to complex structures such as the vertebrate eye or ear. Sensory organs can concentrate and amplify a sensory signal. The bones of the human inner ear amplify sound waves by 20 times before they reach the receptors. Sensory organs can also allow an animal to detect which direction a sensory signal is coming from.

Sensing conditions within the body

The sensory systems in vertebrates send information to the nervous system, which then adjusts conditions, usually without conscious control. Mechanoreceptors, for example, are sensory neurons that detect the degree of muscle contraction and the position of the skeletal system. The nervous system processes the information and sends signals to the

muscles to maintain balance when we are standing or walking. Sometimes, we are more aware of the conditions within the body. For example, sensory receptors in the mammalian brain may detect an increasing concentration of substances such as salt and sugar in the bloodstream. This signal will be processed by the brain, which then stimulates a sense of thirst. After the mammal has a drink of water, the concentration of salt and sugar in the blood decreases and the feeling of thirst diminishes.

IN FOCUS

Sensory transduction

The method by which a receptor converts a sensory input to an electrical signal is called sensory transduction. For example, the sense of smell relies on specific chemicals binding with protein molecules in the cell membrane of an olfactory receptor cell. The binding causes ion channels in the membrane to open, allowing electrically charged ions into, or out of, the receptor cell. The movement of charged ions alters the receptor cell's resting potential, which can initiate action potentials. These impulses then travel along nerves to the brain, where the relevant smell is perceived.

Cephalopod and vertebrate eyes

The visual system is extremely useful for most animals, since it allows a great deal of information to be detected, often at long distances. For this reason, eyes are found in several groups. Eyes in cephalopods, such as octopuses and squid, and vertebrates are remarkably similar in many ways. Both types of eyes are spherical, with an outer layer (cornea) and a small area where light can enter (the pupil). In both cases, the diameter of the pupil varies according to the light intensity. Both types also have a lens to help focus the light onto a sensory surface (the retina). In each case, the eye has a similar function and needs the same high performance levels. The similarity of eyes in cephalopods and vertebrates is an example of convergent evolution—structures with different origins that have evolved to become alike.

Sensing the outside world

Nervous systems can extract a wide variety of information about the outside world. The types of information can be broadly divided by the form of input that is being detected.

Mechanoreceptors detect mechanical energy. This mechanical sense can be direct, such as touch and pressure; or indirect, as when an ear detects high-frequency vibrations in the air that are perceived as sound. The sense of balance

▼ EYE
Human
Light rays are focused by the cornea and lens to produce an upside-down image on the retina, the light-sensitive tissue lining the back of the eye. Electrical signals from stimulated cells in the retina travel to the brain via the optic nerve for interpretation.

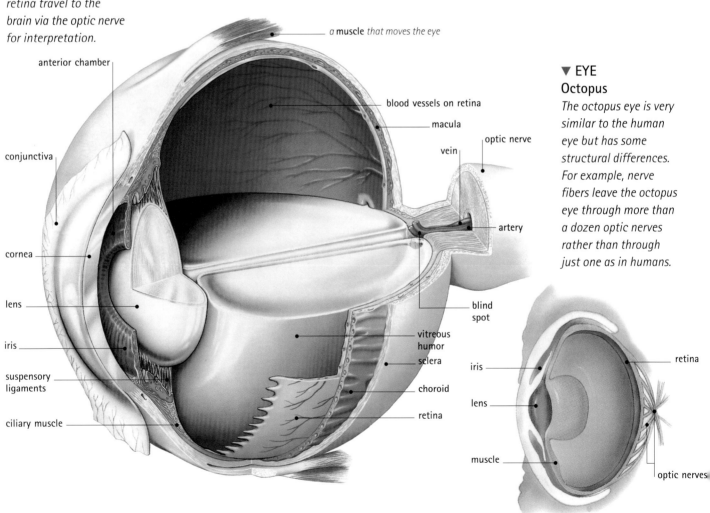

a **muscle** *that moves the eye*

anterior chamber

blood vessels on retina

macula

optic nerve

vein

conjunctiva

cornea

artery

lens

blind spot

iris

vitreous humor

suspensory ligaments

sclera

choroid

ciliary muscle

retina

▼ EYE
Octopus
The octopus eye is very similar to the human eye but has some structural differences. For example, nerve fibers leave the octopus eye through more than a dozen optic nerves rather than through just one as in humans.

iris

retina

lens

muscle

optic nerves

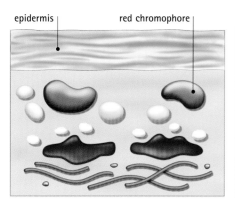

epidermis | red chromophore

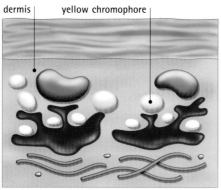

dermis | yellow chromophore

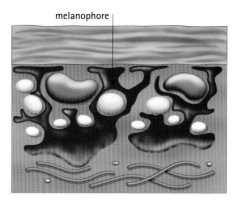

melanophore

▲ **COLOR CHANGE IN LIZARDS**
Chameleon, gecko, or agama
In some lizards, the ability to change skin color is controlled by the nervous system.

When red and yellow pigment cells called chromatophores dominate near the surface of the dermis, the lizard appears orange. As black and brown cells called melanophores

move up the dermis, the lizard appears terra-cotta. As the melanophores spread out under the epidermis, the lizard darkens to chocolate brown.

in mammals is also a form of mechanoreception as it relies on the movement (due to gravity) of hairlike projections on receptors in the vestibular apparatus of the inner ear.

Nociceptors detect pain. Different types of nociceptor respond to different causes of pain: extreme heat, pressure, chemicals (such as acid), or inflammation. Pain is an important sense, since it can result in avoidance of a damaging contact. Also, a painful damaged part of the body is more likely to be rested, speeding healing.

Photoreceptors detect light. The sensory range of visual systems can be very large in different animals and is often well outside the human visual spectrum. Insects, birds, and fish can detect ultraviolet light (wavelengths that are shorter than 400 nm). Some snakes have

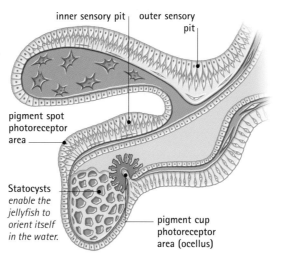

inner sensory pit | outer sensory pit

pigment spot photoreceptor area

Statocysts *enable the jellyfish to orient itself in the water.*

pigment cup photoreceptor area (ocellus)

◄ **NERVOUS SYSTEM**
Jellyfish
Around the edge of a jellyfish's bell are centers called rhopalia, which contain various sensory areas that enable the animal to detect light and chemicals and to detect the orientation of the bell with respect to gravity.

EVOLUTION

Losing your senses

Animals that are not exposed to a certain type of sensory input have often evolved over millennia in such a way that they lose the ability to detect this input. Dispensing with an unnecessary sense organ saves metabolic energy. This can be seen in certain animals, including some fish, that live in permanent darkness in caves and have lost their ability to see. Extreme cases can be seen in adult parasites, such as the tapeworm, that live inside warm-blooded animals. In these circumstances, the conditions are so constant that the parasites have very few sensory structures.

▶ *The yellow clumps in this colored micrograph of the human inner ear are hairlike stereocilia, which project from the ends of sensory cells. The inner ear converts sound waves into nerve impulses by stimulating the stereocilia, and these impulses travel to the brain for interpretation. Magnified 2,500 times.*

Hearing with legs and tasting with feet

Exactly where particular types of sensory information are detected varies in different animals. Sound is very important for crickets—for example, in finding a mate. To tell where a sound is coming from, the "ears" need to be as far apart as possible. The farthest distance possible on a small insect is between the legs, so crickets have tympanic membranes on their legs, which they use to hear mating calls. Flies need to detect food substances when they first land on a surface, so they have chemoreceptors on, and can "taste" with, their feet.

Chemoreceptors can detect both simple and complex chemicals. For example, the ability to taste salt (sodium chloride) is brought about by chemoreceptors on the tongue. Olfaction, the sense of smell, is another form of chemoreception. Most chemoreceptors respond to general groups of structurally similar molecules. Sometimes the receptor is highly sensitive to just one chemical of particular importance to an animal. Male moths often have chemoreceptors in their antennae, which are particularly sensitive to sex pheromones released by female moths.

Electroreceptors can detect electrical fields and exist in animals as diverse as sharks and platypuses. Electroreceptors are used to detect the small electrical fields around prey items buried in sand and mud, where other senses cannot penetrate.

There is good evidence to suggest that some animals (such as pigeons) can detect the earth's

receptors capable of detecting infrared radiation, which they use to detect the body heat of their prey. Honeybees can see the polarization of light in the sky, and they use it to navigate. For many animals, the way in which they perceive the environment must be very different from a human's point of view.

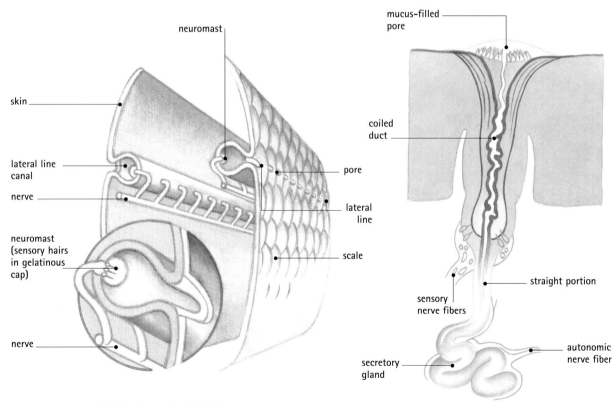

▲ LATERAL LINE SYSTEM
Trout
The lateral line system is a sensory organ that runs around the head and down the sides of a fish. Sensory hairs in structures called neuromasts detect movements in water, allowing the fish to sense not only its prey but also its predators.

▲ ELECTRORECEPTOR
Platypus
There are about 10,000 mucus-filled electroreceptor glands in the platypus's bill. These receptors help the animal locate prey underwater or even buried in mud where they cannot be seen.

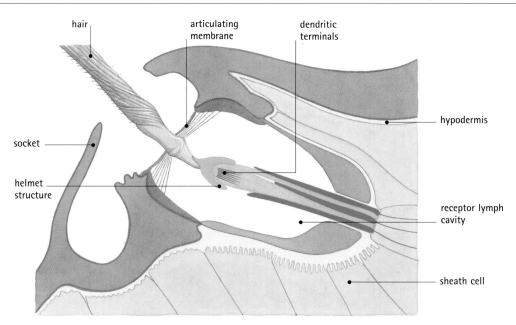

hair

articulating membrane

dendritic terminals

▶ TRICHOBOTHRIUM
Spider

Trichobothria are long, fine hairs arising from sockets on the spiders' legs. These hairs are mechanoreceptors, which detect air vibrations and currents.

socket

helmet structure

hypodermis

receptor lymph cavity

sheath cell

magnetic field and use it for navigation. The brains of pigeons contain magnetite, a naturally magnetic mineral. However the exact location, anatomy, and physiology of the possible magnetoreceptors have yet to be found.

Motor systems: Effectors

The output of the nervous system travels to organs called effectors. By far the most common types of effectors are muscle cells of various types. Motor neurons make synapses with muscle cells; and when the neurons are active, the muscle contracts. In addition to muscles, there are other types of effectors. One main group comprises the hormone-producing glands of the endocrine and exocrine systems, in which neuronal activity can stimulate or stop the release of substances. For example, in mammals the hypothalamus, at the base of the brain, can signal the pituitary gland to release a variety of hormones into the bloodstream.

▼ *This light micrograph shows a neuromuscular junction, where a nerve cell connects to muscle. Four ends of an axon branch onto muscle cells, terminating in synapses (black dots). Magnified 300 times.*

sclera

choroid layer

pigment layer

rod

cone

horizontal cell

bipolar cell

amacrine cell

ganglion cell

nerve fibers of optic nerve

▲ CROSS SECTION OF RETINA
Mammalian

Rods and cones produce electrical signals when light strikes the retina. These signals are modified by other cell types, such as bipolar cells, before leaving the retina via the optic nerve.

Invertebrate nervous systems

Invertebrates possess a dazzling array of different nervous systems. Even within a single taxonomic group, there is much variation. This can be demonstrated by looking at the mollusks. Some mollusks, such as clams, have just a few widely spaced and similarly sized ganglia. No specific coordinating brain is needed for an adult life spent in one place. Octopuses, on the other hand, have a very sophisticated nervous system with a relatively large brain. This complexity reflects their active hunting behavior and excellent learning and memory abilities.

Nerve nets to nerve cords
Among the simplest nervous systems is that of cnidarians called hydras. (Cnidarians include corals, sea anemones, and jellyfish.) A hydra's neurons are arranged in a netlike manner. Since these nerve nets lack ganglia, there is no central control area, and the processing of information is distributed throughout the nervous system.

Most of the synapses between neurons are electrical, and action potentials can travel in either direction along the axon. This structure enables a stimulus at any point to trigger action potentials throughout the nerve net and lead to a generalized response. Other cnidarians show some degree of grouping of neurons, particularly around the mouth, where sensory information is most varied. Some planarians (flatworms) possess a network of nerves throughout the body, and they have a distinct brain in the head region. Other planarians have two nerve cords extending along the body, connected and extended by a series of lateral nerves to form a ladderlike system.

Complex invertebrate nervous systems
Even an animal as apparently simple as the earthworm (an annelid) has a complex nervous system. Earthworms possess a brain and a nerve cord along the body, with a ganglion in each segment. Each ganglion sends lateral nerves

▶ **Giant clam**

The giant clam does not have a true brain. Instead, it has three pairs of interconnected ganglia. These are the cerebropleural ganglia, which send nerve fibers to the the palps (feeding appendages), mantle (body), and anterior adductor muscle; the visceral ganglia, which send nerve fibers to the heart, mantle, posterior adductor muscle, gut, gills, and siphon; and the pedal ganglia, which send nerve fibers to the foot.

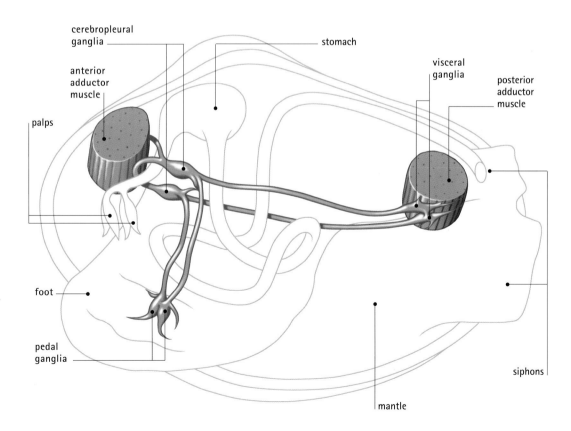

cerebropleural ganglia

stomach

visceral ganglia

anterior adductor muscle

posterior adductor muscle

palps

foot

pedal ganglia

siphons

mantle

throughout the segment to gather sensory information and distribute motor commands. This basic pattern is also found in many other invertebrate groups, including the arthropods (such as insects and crustaceans). In arthropods, the segmental ganglia have often evolved to be physically fused, so their nervous system is more centralized and specialization of groups of segments is possible.

A very different nervous system is seen in the echinoderms (starfish and sea urchins). These have a radial nervous system with nothing resembling a central brain. There is a nerve ring around the mouth and nerves extending into each arm. Sometimes there are small ganglia associated with the tube feet. The unique nervous system of echinoderms allows any radial portion of the body to act as the "head." This feature also allows the nervous system to be regenerated following even major injury, such as the loss of a limb in a starfish.

▶ **Earthworm**
This worm has a brain and a nerve cord running along its body. Within each segment, a ganglion gives rise to three main nerves (one is shown here), which divide further.

COMPARATIVE ANATOMY

Naming neurons

Some invertebrate species have specific neurons that can be identified from one individual animal to the next. The neurons can then be named (or numbered) and their functions studied. Some of the specific neurons control particular aspects of a behavior. For example, artificially stimulating action potentials in a neuron called CBI-2 in the sea hare (a type of sea slug) induces feeding rhythms—as though the animal is eating. In most vertebrates, the neurons are so small and numerous that this kind of identification and naming is not possible. One exception is the Mauthner cell in bony fish, which induces an escape response when it is stimulated.

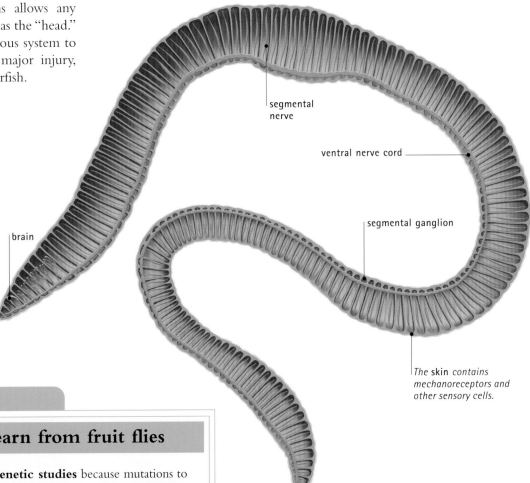

segmental nerve

ventral nerve cord

segmental ganglion

brain

*The **skin** contains mechanoreceptors and other sensory cells.*

GENETICS

What we can learn from fruit flies

Fruit flies are used in many genetic studies because mutations to individual genes are relatively easy to induce and isolate. Several of these mutations have involved changes to nervous system function. Mutant fruit-fly genes called dunce and rutabaga result in poor learning. The mutations in these cases affect enzymes involved in sending messages within neurons. Neuroscientists can use these mutations to study how learning is achieved at the cellular level.

Vertebrate nervous systems

Vertebrate nervous systems are not as varied as those of invertebrates. However, vertebrate nervous systems are generally capable of much higher levels of processing and interpreting information. Vertebrate nervous systems also typically generate more flexible and sophisticated behaviors. This is particularly so in birds and mammals and especially in the human nervous system. The vertebrate nervous system can be broadly divided into two parts: the peripheral nervous system (PNS) and the central nervous system (CNS). The CNS consists of the brain and spinal cord, and the PNS is made up of the nerves connecting the CNS with the rest of the body, including the sense organs and muscles.

Peripheral nervous system

Anatomically, the PNS consists of a series of nerves extending from the brain (the cranial nerves) and spinal cord (the spinal nerves). The PNS has two main functions. It collects sensory information and distributes motor commands. Usually, each nerve contains both motor and sensory axons. Exceptions are the optic and olfactory nerves, which only collect sensory information. The PNS can be further divided into sections called the somatic system and the autonomic system. The somatic system receives information from the external senses (such as touch and vision) and sends motor commands to the skeletal muscles. This system deals with reflexes and voluntary movements.

▶ **BRAIN**
Human
The human brain consists of the brain stem, the cerebellum, and the cerebrum, which has four lobes. Body activities are controlled by specific areas within the brain.

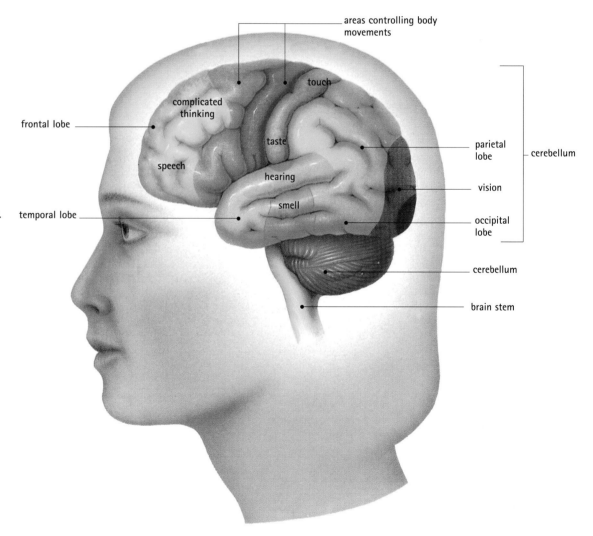

areas controlling body movements

complicated thinking

touch

frontal lobe

taste

speech

hearing

smell

temporal lobe

parietal lobe

cerebellum

vision

occipital lobe

cerebellum

brain stem

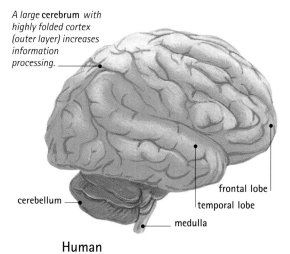

*A large **cerebrum** with highly folded cortex (outer layer) increases information processing.*

cerebellum

frontal lobe

temporal lobe

medulla

Human

◄ COMPARING BRAINS

This diagram shows the typical brain structures of a human, a bird, a reptile, and a fish. All vertebrate brains have certain features in common, including a cerebellum to coordinate movement, optic lobes for sight, and olfactory lobes to interpret smells.

Get rhythm

Some behaviors, such as eating, breathing, and locomotion, are rhythmic. In many of these cases, there is a special network of neurons that produces a basic rhythmic neuronal output. These networks are called central pattern generators. In most cases, this basic rhythm has to be started, stopped, and fine-tuned by ongoing sensory input. For example, a basic locomotor rhythm is generated in the spinal cord of vertebrates and then altered by sensory feedback and input from higher centers in the brain to produce walking or running, depending on the circumstances.

Bird

*The pronounced **cerebrum** is concerned with flight control.*

medulla optic lobe pituitary olfactory lobe

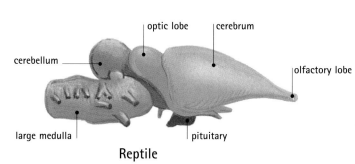

optic lobe cerebrum

cerebellum

olfactory lobe

large medulla pituitary

Reptile

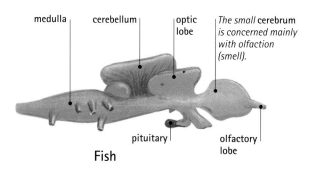

medulla cerebellum optic lobe

*The small **cerebrum** is concerned mainly with olfaction (smell).*

pituitary olfactory lobe

Fish

The blood–brain barrier

The circulatory system plays a vital role, bringing oxygen and nutrients to the brain. However, the brain, more than any other organ, must also be protected from changes in concentrations of substances and from infections. This is accomplished by a layer of brain endothelial (lining) cells that are connected to each other very tightly. This layer is called the blood–brain barrier. The blood–brain barrier forms a physical barrier to some molecules, bacteria, and viruses. The barrier also regulates which substances can enter or leave the brain. When the blood–brain barrier is breached, as in bacterial meningitis, the results can be very serious or even fatal. However, the blood–brain barrier also stops many medicines from reaching the brain, so drug molecules have to be designed very carefully.

The autonomic system receives information about the internal state of the body and then sends commands to maintain these conditions at appropriate levels. The responses controlled by the autonomic system are usually not under voluntary control. Examples are the size of the pupil of the eye, salivation, heart rate, and control of the digestive and excretory systems. Opposite aspects of each response are controlled by separate parts of the autonomic system. These are the sympathetic and parasympathetic systems. The sympathetic system, for example, controls dilation of the pupil of the eye, and the parasympathetic system controls constriction of the pupil.

▶ CEREBRAL LOBES
Human

The cerebral cortex of the human brain is divided into four lobes: frontal, temporal, parietal, and occipital. Each lobe has distinct functions, some of which are shown in this diagram.

▼ SPINAL CORD
Human

This diagram shows paired spinal nerves leaving the spinal cord. Sensory nerves enter the spinal cord via the dorsal root, and motor nerves exit by way of the ventral root. The sympathetic chain of ganglia runs parallel to the spinal cord.

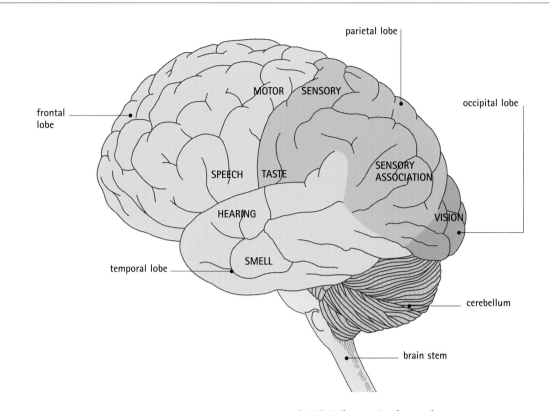

EVOLUTION

Ancient bird brains

Recent studies have shown that one of the first birds to evolve, archaeopteryx, had a brain comparable in size to that of modern birds and three times larger than the brains of similar-size dinosaurs. Scientists believe archaeopteryx had extra brain power that enabled it to deal with all the sensory processing and muscular coordination needed for flight. Evolution of the CNS was therefore just as necessary as the evolution of mechanical flight systems for flying to occur in birds.

CNS: The spinal cord

The spinal cord extends down the back of vertebrates and is protected by vertebrae, which are the bones that make up the spine. The major function of the spinal cord is to carry information to and from the brain. The spinal cord can also generate the basic motor rhythms for breathing and locomotion. It also provides some relatively simple responses, such as reflexes. A reflex is an automatic response to a specific stimulus that can occur without sending or receiving any information to the brain. In the knee-jerk reflex, the leg is struck just below the knee. This stretches a tendon and activates sensory neurons that directly synapse with, and activate, motor neurons. These, in turn, cause muscles to contract, jerking the lower leg forward. All this happens without the need for any more complex integration. Most reflexes also involve spinal interneurons between the sensory and motor neurons.

CNS: The hindbrain and midbrain

The rearmost section of the brain is called the hindbrain. That is where the brain and spinal cord connect. Just in front of this junction is a region of the hindbrain called the medulla oblongata, or medulla. The medulla coordinates several autonomic functions, including digestion, breathing, vomiting, and

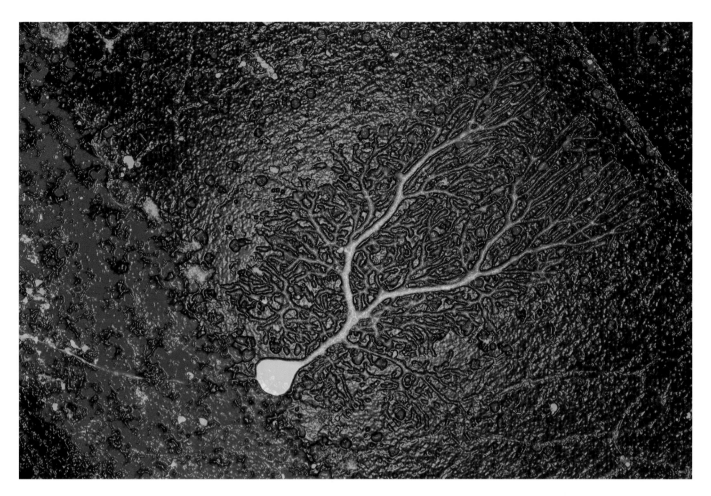

swallowing. The other major section of the hindbrain is the cerebellum (named for the Latin "little brain"). This is a domelike structure behind, and in the case of mammals located beneath, the cerebrum (part of the forebrain). The cerebellum receives sensory information about the position of the body, as well as input from the visual and auditory systems. In addition, higher brain centers in the cerebrum relay information about motor commands to the cerebellum, which uses this information to maintain balance and provide automatic coordination of movements.

In front of the hindbrain is the midbrain. Overall, the midbrain integrates sensory information, especially vision and hearing. Specific areas deal with each of these: the inferior colliculi for hearing and the superior colliculi for vision. In mammals, however, visual information is dealt with mostly by the cerebrum, and the role of the superior colliculi is reduced. The reticular formation regulates the state of arousal in vertebrates. Arousal is the amount of interest in—or

awareness of—the environment. For example, the lowest state of arousal is sleep, and the highest occurs when predators are sensed.

CNS: The forebrain

The most complex neuronal processing is accomplished by the forebrain. This consists of a number of areas associated with different functions. The thalamus receives and sorts many types of sensory information and

▲ *This micrograph shows a Purkinje nerve cell. Large numbers of these highly branched cells occur in the outer layer of the cerebellum. The axons (nerve fibers) of Purkinje cells carry information processed in the cerebellum to the brain stem.*

EVOLUTION

Evolution of the vertebrate nervous system

Increasing size is one major trend in the evolution of the vertebrate brain. Among fish, amphibians, and reptiles, the size of the brain is broadly constant, relative to body size. In birds and mammals, the most recently evolved vertebrates, the brain is comparatively large. A small mammal will have a much bigger brain compared with a reptile of the same size. The cerebrum, in particular, is larger in birds and mammals; in some mammals, the outer layer (cortex) is folded, which increases information processing.

▶ CEREBRAL HEMISPHERES
Human

The cerebrum has two halves, or hemispheres, which process information from opposite parts of the body. For example, when the eyes are fixed on a point straight ahead, information from the left field of vision of each eye is sent to the right cerebral hemisphere, and information from the right field of each eye is sent to the left cerebral hemisphere.

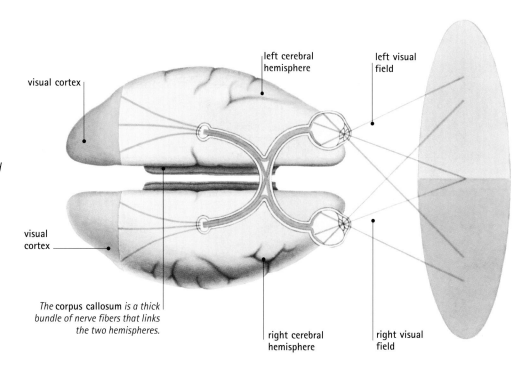

visual cortex

left cerebral hemisphere

left visual field

visual cortex

*The **corpus callosum** is a thick bundle of nerve fibers that links the two hemispheres.*

right cerebral hemisphere

right visual field

IN FOCUS

Crossed wires

The axons descending from the brain to the spinal cord cross sides as they pass through the hindbrain. As a result of this arrangement, motor commands from the left side of the brain actually go to the right side of the body. The visual system is set up in this way, too, so information from your right eye ends up going to, and being processed by, the left side of your brain. The region where nerve fibers cross hemispheres, and hold the two hemispheres together, is the corpus callosum.

then relays it on to higher brain regions for further processing and interpretation. The hypothalamus is vital in maintaining the internal body state at optimal conditions. It regulates body temperature, hunger, thirst, and sexual functions. The limbic system is a group of structures that influence basic emotional responses. Two parts of the limbic system, the hippocampus and amygdala, are thought to be important in learning and memory.

The frontal part of the vertebrate brain is called the cerebrum. In fish and amphibians, this is relatively small and mostly concerned with processing olfactory (smell) information. In many mammals, the cerebrum is the largest part

◀ SENSORY CORTEX

The surface of the cerebrum, the cortex, has areas that receive incoming sensory information and other areas that are concerned with outgoing motor information. The different parts of the body have different amounts of the cortex devoted to them. The diagram shown is a map of the sensory and motor cortexes, proportional to how much of the cortex is devoted to each body part.

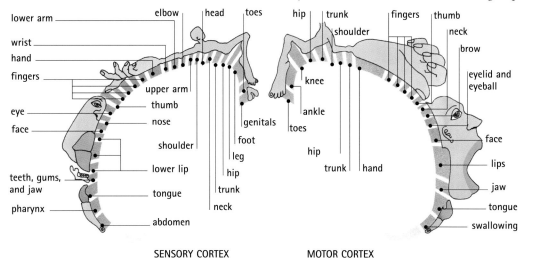

SENSORY CORTEX

MOTOR CORTEX

of the brain and has many folds. These folds, or convolutions, increase the surface area of the outer layer, which is the cerebral cortex. Since the neuronal cell bodies are all located in the cerebral cortex, a greater surface area allows more neurons and therefore greater processing power.

The cerebrum is the largest and most conspicuous feature of the human brain, with many large convolutions. The human cerebrum consists of left and right cerebral hemispheres, connected by a nerve tract called the corpus callosum. The functions of different regions of the cerebrum have been mapped to some extent using various imaging techniques. Two examples are the motor cortex and sensory cortex. Motor output and sensory information (mainly touch) from different parts of the body can be mapped out across these cortices. This mapping is not uniform, because different body regions vary in their relative importance. For example, the areas in the motor cortex devoted to hands and face are larger than those for the entire torso and legs. Other specific regions of the cerebrum are devoted to processing sensory input from the remaining senses: smell, taste, hearing, and vision. Further regions are active during particular activities, such as reading, speech, and hearing.

Higher brain functions

The higher brain functions most commonly associated with humans, such as memory, emotion, speech, and thought, have been—and continue to be—the subject of extensive scientific investigations. The anatomical centers associated with some of these processes are known, but scientists still have little knowledge of their detailed working.

The different hemispheres of the brain control different functions. For example, the left hemisphere controls speech, language, and calculation, whereas the right hemisphere controls spacial awareness and creative thinking. The processing of separate aspects of complex abilities, such as language and speech, is often distributed around the brain. For example, different sections of the brain are active when someone is listening to a word, thinking of a word, or saying a word. These regions must interact in complicated and fast, yet subtle, ways, even in the most simple conversation between humans.

IN FOCUS

Imaging living nervous systems

There are several modern ways of imaging the nervous system, in addition to conventional X rays. Positron emission tomography (PET) measures positrons emitted by minute amounts of a radioactive substance administered to a patient. PET scans can be used to measure levels of metabolic activity or blood flow to different areas of the brain. These indicate which parts of the brain are most active at any time. Computed tomography (CT) uses X rays to generate a series of images of "slices" through the head. These can then be built up into a three-dimensional image. An electroencephalogram (EEG) uses electrodes attached to the surface of the head to map the electrical activity in different parts of the brain.

▼ NORMAL BRAIN
This PET scan of a human brain shows a large area (brown) of normal activity.

▼ BRAIN WITH DEMENTIA
In someone with the degenerative brain disorder senile dementia, there is reduced brain activity.

large area of activity   reduced area of activity

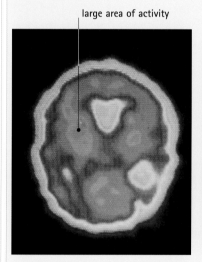

The nervous system enables humans to accumulate and use memories and have creative and imaginative thought processes and makes us aware of our surroundings, ourselves, and each other. It also enables humans to communicate complex ideas and strive to understand them. A fuller understanding of these processes is one of most challenging and interesting fields in science today.

RAY PERRINS

FURTHER READING AND RESEARCH

Freeman, Scott. 2001. *Biological Science*. Prentice Hall Publishers: Saddle River, NJ.

Newt

CLASS: Amphibia ORDER: Urodela
FAMILY: Salamandridae

Newts and their close relatives, salamanders, are tailed amphibians. Together, newts and salamanders make up the order Urodela, one of three orders of amphibians. There are about 550 species of newts and salamanders, and they are divided into 10 families. All the species usually called newts belong to the family Salamandridae, along with the European salamanders.

Anatomy and taxonomy

The family Salamandridae contains 74 species in 20 genera. Most live in Europe and Asia, with six species in North America and three in coastal regions of North Africa. Some species are very common and widespread, but others are rare. One species, the Kunming Lake fire-bellied newt, was abundant in Yunnan, China, as recently as the 1950s, but has not been seen since 1979; it is probably extinct.

● **Animals** All animals are complex, multicellular organisms that depend on plants or other animals for food. Most are capable of movement and have sense organs that enable them to detect external stimuli and respond if necessary.

● **Chordates** Chordates have, at some stage during their development, a stiff supporting rod called a notochord along their back.

● **Vertebrates** Vertebrates are animals with a backbone, or spine, consisting of connected bones called vertebrae. The head, trunk, and limbs are attached to the backbone by a series of muscles that control movement.

● **Amphibians** The amphibians are vertebrates that have a moist, glandular skin. They usually lay their eggs in water or in some other moist location. They may fertilize their eggs internally or externally, but male amphibians do not have a specialized organ for copulating. Eggs hatch into aquatic larvae called tadpoles,

which eventually change into the adult form. A small number of species give birth to live young, whereas other amphibians retain their larval form throughout life. Some amphibians remain in water as adults. There are three orders of amphibians: salamanders (order Urodela); frogs and toads (order Anura); and caecilians (order Gymnophiona). Most amphibians are easy to recognize, with the possible exception of the limbless caecilians, which superficially resemble earthworms or eels.

● **Salamanders** Almost uniquely among the amphibians, salamanders have legs and a tail in the adult stage of their life. Some caecilians have a very short tail but no legs, and adult frogs and toads have legs but—with the exception of two species—no tail. There are 10 families of salamanders.

● **Sirens** The four species of sirens (family Sirenidae) are aquatic salamanders that live in North America. They have forelimbs, but no hind limbs, and they retain larval gills throughout their life.

▼ *The family tree shows the relationship of the three main groups of amphibians and some of the families of newts and salamanders mentioned in the text.*

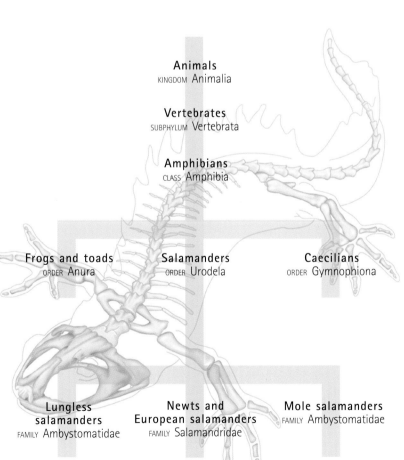

Animals
KINGDOM Animalia

Vertebrates
SUBPHYLUM Vertebrata

Amphibians
CLASS Amphibia

Frogs and toads
ORDER Anura

Salamanders
ORDER Urodela

Caecilians
ORDER Gymnophiona

Lungless salamanders
FAMILY Ambystomatidae

Newts and European salamanders
FAMILY Salamandridae

Mole salamanders
FAMILY Ambystomatidae

● **Hellbenders and giant salamanders** There are three species in the family Cryptobranchidae: the hellbender lives in North America; the Chinese giant salamander inhabits parts of China; and the Japanese giant salamander is a Japanese species. All are aquatic. Scientists believe that since they fertilize their eggs externally, giant salamanders appeared early in the history of salamander evolution.

● **Asiatic land salamanders** These salamanders (family Hynobiidae) inhabit much of Central and East Asia, including Japan. They are mostly terrestrial, except when breeding, and practice external fertilization.

● **Mud puppies and relatives** These aquatic salamanders (family Proteidae) retain larval gills. Five species live in the southeastern United States, and one (the olm) inhabits underground rivers in southeastern Europe.

● **Torrent salamanders** Torrent salamanders (family Rhyacotritonidae) occur in the Pacific Northwest of North America. The four small species favor cool forests with seeps or fast-flowing streams in which their larvae develop.

● **Amphiumas** Amphiumas (family Amphiumidae) have very tiny limbs and are eel-like in appearance. They are completely aquatic and live only in the southeastern United States.

● **Lungless salamanders** There are about 380 species of lungless salamanders (family Plethodontidae), making them the largest family of salamanders. Most live in North

▲ *The northern red salamander has no lungs, and respiration takes place only through its skin. This species is a member of the family Plethodontidae and lives in North America.*

and South America and parts of Europe, but one species occurs in Korea. They lack lungs; all their respiration takes place across their skin. Several species of lungless salamanders retain their larval form throughout life (they are neotenic). Such forms live both in surface streams and in cave waters. The majority of plethodontids are terrestrial and live in damp, rocky, or grassy places in woodlands or in caves. Some, especially tropical species, live in trees.

● **Mole salamanders** Mole salamanders, which make up the family Ambystomatidae, occur in North America. There are 30 or so species, all in the genus *Ambystoma*, and they are heavy-bodied salamanders that live in a variety of habitats. The tiger salamander is the most widespread and familiar species. Some species, called axolotls, retain larval characteristics throughout their life.

● **Pacific giant salamanders** These robust salamanders (family Dicamptodontidae) live in western North America. They live as stream-dwelling larvae for several years, and some achieve sexual maturity as larval forms.

● **Newts and European salamanders** The 74 species in this group (family Salamandridae) live in Europe, Asia, and North America. Most are land dwellers that return to water to breed. A few, however, breed on land, and some give birth to fully formed juveniles, skipping the free-living larval stage altogether. The 12 species of typical newts live in Europe and parts of western Asia. Typical newts breed in ponds and lakes, where the females attach eggs individually to the leaves of aquatic plants. The color of the males intensifies in the breeding season, and in many species the males also develop features such as a dorsal crest, webbed feet, or a tail filament.

FEATURED SYSTEMS

EXTERNAL ANATOMY Newts are amphibians with four limbs and a tail. They have soft, moist skin containing many glands. *See pages 802–804.*

SKELETAL AND MUSCULAR SYSTEMS Newts have a conventional, if rather simple, tetrapod skeleton, with a wide, flattened skull, backbone, short ribs, and limb girdles with paired limbs. *See pages 805–806.*

NERVOUS SYSTEM Smell is the most important sense in most species of newts and salamanders, although they also use vision to detect prey. Their hearing is poor. *See page 807.*

CIRCULATORY AND RESPIRATORY SYSTEMS Newts have lungs but can also exchange gases through the skin and the lining of their mouth. The larvae breathe with gills. *See pages 808–809.*

REPRODUCTIVE SYSTEM Newts and salamanders practice external fertilization and often have elaborate courtship rituals that help them maximize the chances of breeding successfully. *See pages 810–811.*

External anatomy

CONNECTIONS

COMPARE the moist skin of a **NEWT** with the scaly skin of a reptile such as a **JACKSON'S CHAMELEON** or a **GREEN ANACONDA**. Amphibians depend on water for breeding and must live in a humid environment. Reptiles have a waterproof skin, have more efficient lungs, and lay shelled eggs that do not dry out. Reptiles' scales reduce water loss while maintaining flexibility, and they provide protection from rough surfaces.

Newts and salamanders belonging to the family Salamandridae have four relatively long limbs and a tail. Some other species of salamanders have very small limbs. Newts' adult and larval forms are distinct from each other, with the larvae being recognizable by their frilly external gills. The larvae of crested newts are called pond-type larvae because they have a stout body and a high dorsal fin that reaches almost to the head. Other newts may have stream-type larvae: they are more streamlined, enabling them to resist the flow of running water and not get swept downstream.

When they hatch, crested newt tadpoles are still at an early stage of development. They have small front limb buds, and their mouth is not fully formed. The tadpoles have a pair of balancers projecting down from their head that they use to cling to an empty egg capsule or a piece of aquatic vegetation. The tadpoles remain in that position for a few days, while they feed on the yolk supply. When the yolk is used up, the tadpoles reabsorb the balancers, and as the mouth and legs become functional, the tadpoles go in search of food. The front legs grow, the hind legs appear, and the larvae move

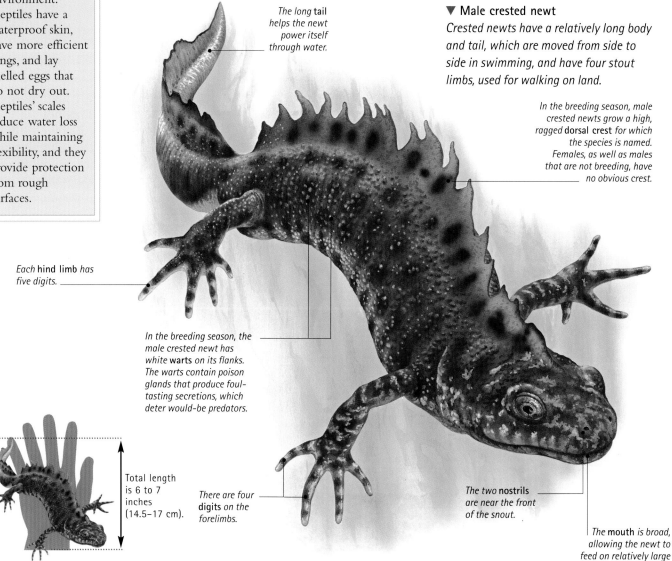

The long **tail** helps the newt power itself through water.

▼ Male crested newt
Crested newts have a relatively long body and tail, which are moved from side to side in swimming, and have four stout limbs, used for walking on land.

In the breeding season, male crested newts grow a high, ragged **dorsal crest** for which the species is named. Females, as well as males that are not breeding, have no obvious crest.

Each **hind limb** has five digits.

In the breeding season, the male crested newt has white **warts** on its flanks. The warts contain poison glands that produce foul-tasting secretions, which deter would-be predators.

Total length is 6 to 7 inches (14.5–17 cm).

There are four **digits** on the forelimbs.

The two **nostrils** are near the front of the snout.

The **mouth** is broad, allowing the newt to feed on relatively large prey: worms, slugs, tadpoles, and frogs' eggs.

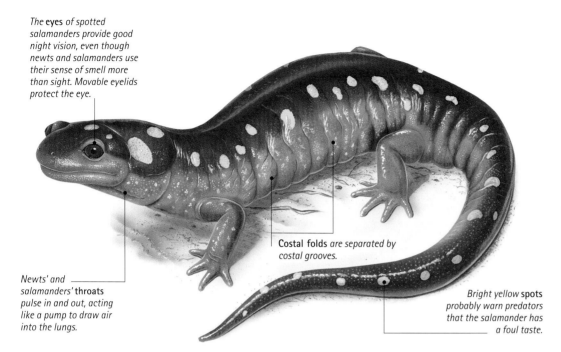

*The **eyes** of spotted salamanders provide good night vision, even though newts and salamanders use their sense of smell more than sight. Movable eyelids protect the eye.*

*Newts' and salamanders' **throats** pulse in and out, acting like a pump to draw air into the lungs.*

Costal folds *are separated by costal grooves.*

*Bright yellow **spots** probably warn predators that the salamander has a foul taste.*

◀ **Spotted salamander**
Compare the smooth skin of this relative of the newt with the rough skin of the crested newt. This species is 6 to 7 inches (15–18 cm) long when adult. It has a round jaw, a robust body, and strong legs. The folds on its flanks (costal folds) correspond to groups of muscles. The spotted salamander is a member of the family Ambystomatidae (the mole salamanders).

by walking on the bottom of the pond or swimming in short bursts. The large, frilly, external gills help it extract as much oxygen as possible from the water, and the high fins increase the surface area of skin across which oxygen can be absorbed.

Metamorphosis begins when the larva is fully grown. As the lungs develop and the external gills are reabsorbed, the larva must make frequent journeys to the surface to breathe. Pigment begins to color the larva's body until it looks like a smaller version of its parents. Then, it usually leaves the water to begin a terrestrial life.

In a few species, such as the alpine salamander and some populations of the fire salamander, the larvae develop inside the mother and are born as fully developed juveniles. In two families of totally aquatic salamanders (sirens and mud puppies) adults retain juvenile characteristics throughout their life, a phenomenon called neoteny.

Adult newts that spend much of their time in the water often have a tail that is flattened from side to side. This tail shape helps newts swim more efficiently. Salamanders that remain on land have a cylindrical tail. The tall fins of male crested salamanders are not rigid enough to enhance the paddle shape of the tail. Male aquatic adults may also have webbed feet and dorsal crests during the breeding season. The

crested newt has a high, serrated crest, and other species have a crest that has a straight edge or is wavy. These species reabsorb their crest at the end of the breeding season, when they usually leave the water and return to a land-based lifestyle.

Smooth and rough skin

The skin of newts may be smooth or rough. Crested newts' skin is very rough and warty, so the animal is sometimes given the alternative name "warty newt." Newt skin is a complex structure that has several functions. It allows the exchange of gases such as oxygen, thus allowing the newt to breathe through the skin

PREDATOR AND PREY

Spare parts for newts

Newts and salamanders are able to regrow damaged parts. These amphibians are especially prone to mutilation in their larval form, when individuals feed by snapping at any moving thing within range— including each other. It is not unusual to find larvae with missing gills, toes, or limbs. The loss is detected by nerve impulses that stimulate regeneration. Larval newts can regrow a missing toe in a matter of days, although a whole leg takes longer. Adult newts have similar powers, although regeneration takes longer, and the replacement part is usually smaller than the original.

Newt and salamander larval types

Salamander larvae fall into two broad categories. Pond-type larvae have tall tail fins, and the dorsal fin reaches almost to the head. The larvae have a small structure called a balancer on each side of the head during their early stages and long, frilly, external gills. The front limbs are only partially developed, and these larvae have no back limbs—they develop later. Most newts, mole salamanders, and European salamanders have pond-type larvae.

Stream-type larvae have a more flattened body. Their fins are narrow, and the dorsal fin reaches only to the base of the tail. The larvae do not have balancers, and their back legs are developed and functional as soon as they hatch. These characteristics all help larvae that live in clear, fast-flowing, oxygen-rich water. Brook salamanders, torrent salamanders, and many lungless salamanders have stream-type larvae.

(cutaneous respiration), and it also allows water to pass into and out of the newt's body to help maintain water balance.

In the deepest skin layer, called the dermis, there are glands that secrete mucus to keep the skin moist. This feature is important, since gas exchange cannot take place over a dry surface. Other glands produce substances that protect the newt from fungal and bacterial infections, and poison glands produce toxins that protect the newt from predators. In some species, such as the fire salamander, the poison glands are concentrated in certain areas, often just behind the head, where they are called parotoid glands, and along each side of the back. In the crested newt, however, poison glands are scattered all over its body and are responsible for the skin's warty texture. Many predators leave newts alone, finding them distasteful. However, other predators, such as grass snakes, seem unaffected by the toxins and regularly eat newts.

Warning colors

Pigment cells provide additional protection by producing camouflage patterns or warning colors that advertise the newt's poisonous nature. Several species, including crested newts, have a dark-brown or black upper side of the body. The coloration makes them difficult to see if they are resting on the bottom of a muddy pond, for example. The underside of crested newts is brightly marked in orange or red, with bold dark blotches. The coloration warns predators, including fish, which may attack prey from below. If a newt is threatened on land, it arches its back to display the bright warning coloration. This threatening posture is called the *Unkenreflex* because it was first described in the fire-bellied toad, which is called *Unke* in German. In other species of salamanders, the whole animal is brightly colored to warn of highly toxic secretions. In some species, the secretions are potent enough to kill an animal the size of a small dog.

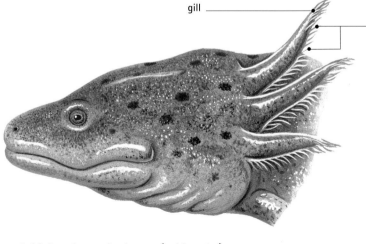

gill — filaments

▲ Mole salamander larvae (cold water)
Mole salamander larvae living in cold water breathe most of their oxygen through the skin. Oxygen dissolves particularly well in cold water. The feathery gill filaments increase the surface area of the skin, thus allowing the larvae to breathe more oxygen.

▼ Mole salamander larvae (warm water)
Oxygen does not dissolve so well in warm water, so mole salamander larvae breathe more actively than their cold-water cousins. By opening the mouth, the larva creates a flow of water through the mouth and out past the gills. The gills absorb oxygen as the water flows past.

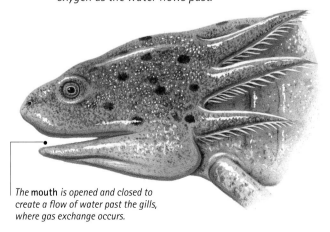

The **mouth** *is opened and closed to create a flow of water past the gills, where gas exchange occurs.*

Skeletal and muscular systems

Newts and salamanders have a simple skeleton, derived from that of the fish from which they evolved. The skull is flattened, and the sockets for the eyes are large. The skull is connected to the backbone in a way that allows it to move up and down, but side-to-side movements are limited; like a person with a stiff neck, a newt must bend its whole body if it wants to look to the side. There are several rows of tiny teeth on the upper and lower jaws and also on the palate (roof of the mouth). All the teeth are a similar shape, and they are used simply for holding onto prey, not for chewing. Newts snap at their food, grip it in their jaws, and then swallow it whole.

The backbone must support the weight of the newt when it is living on land. It does this by transferring the load to the limbs, by means

Sensory systems

When a newt leaves the water, it encounters a problem that arose initially when its ancestors evolved from fish. Since the body is no longer supported by water, the skeleton and muscles have to take the full weight. Thus, movement is slower and more awkward on land than in water. Newts and salamanders are clumsy walkers; they make progress with a side-to-side wriggle, bending the body first one way and then the other and dragging their belly and tail along the ground. Only rarely do newts and salamanders lift their body clear of the ground to move more quickly, as this uses more energy. Small species, with less weight to support, are more agile than larger species, and it is no coincidence that the largest species, such as the giant salamanders, hellbenders, and amphiumas, are completely aquatic.

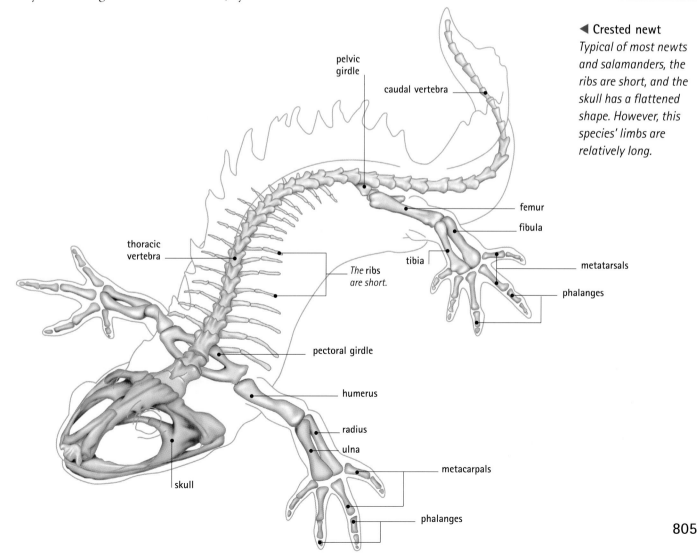

◀ **Crested newt**
Typical of most newts and salamanders, the ribs are short, and the skull has a flattened shape. However, this species' limbs are relatively long.

pelvic girdle

caudal vertebra

femur

fibula

thoracic vertebra

tibia

metatarsals

The ribs are short.

phalanges

pectoral girdle

humerus

radius

ulna

metacarpals

skull

phalanges

▲ *The brown spots along the flank of this ribbed newt mark the points where the tips of the ribs push the skin outward.*

to form a rib cage. The ribs of some species are able to project through the newt's skin, piercing strategically positioned poison glands at the same time. The tail begins immediately behind the pelvic girdle, and in the crested newt and most of its close relatives the tail is roughly the same length as the head and body combined. In some species, such as the gold-striped salamander, the tail is much longer than the body.

of the pectoral and pelvic limb girdles. Newt and salamander limbs are short and stocky. In newts, unlike frogs and toads, the front and hind legs are roughly similar in length. Most newts and salamanders have five digits on the back feet and four on the front, although some species have fewer digits. All salamanders have ribs, but in some species these are very short. When present, the ribs protrude from each side of the backbone without curving around

Muscles

The musculature of newts and salamanders is extremely complex. In species with lungs, hyobranchial muscles move the throat up and down to pump air into the lungs. Other muscles help the animal open and close the mouth, swallow, and move the tongue. Lungless salamanders breathe entirely through their skin and do not possess laryngeal muscles.

Muscles in the trunk enable the salamander to move from side to side. It does this when swimming and walking. The same kind of muscles allow the animal to beat its tail. The musculature of the limbs is responsible for supporting and transferring the body weight during walking, either on land or along the bottom of a pond.

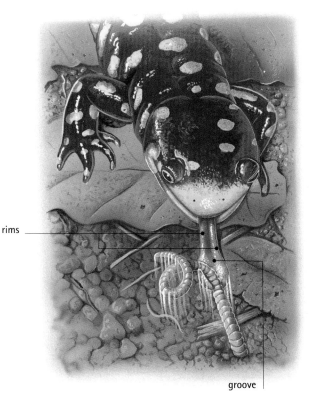

rims

groove

◄ **TONGUE PROJECTION**
Muscles control the movement of a newt or salamander's tongue, which is projected rapidly when prey is within range. Since the center of the tongue forms a groove and the edges form a rim, the tongue engulfs the prey. This tiger salamander is preying on an earthworm.

▲ **TONGUE CONTRACTION**
The tongue is drawn back into the mouth, and then the salamander lurches forward to make sure of a good hold. It shakes its head violently from side to side to kill the earthworm and then swallows it whole.

Nervous system

The central nervous system of newts and salamanders consists of a brain in the skull and a spinal cord running the length of the body. The brain is divided into three sections: the forebrain, midbrain, and hindbrain. The hindbrain merges into the spinal cord, from which nerves radiate out to all parts of the amphibian's body.

Eyes

Newts and salamanders have small eyes. Vision is not as important to them as it is to frogs, although they can detect movement and use their eyes to find food and mating partners. The structure and efficiency of newt and salamander eyes, particularly the shape of the cornea, depend on the environment in which they are used. Purely terrestrial species, such as the fire salamander, have relatively good vision; but species that live partly in the water and partly on land, such as the crested newt, have poorer vision. Experiments with newts have shown that they are farsighted in water and nearsighted on land. Most species, regardless of their natural environment, have good night vision due to the large numbers of rod cells in the retinas of their eyes. Even so, crested newts

and fire salamanders, both of which are highly nocturnal, have a bigger proportion of rod cells (about 50 percent) than species such as the smooth newt, which is active at dusk.

Sound and smell

Newts and salamanders do not have a good hearing system; they do not have external eardrums, and they do not communicate by sound. They are probably not completely deaf, however, since some species react to airborne sounds at certain frequencies.

The sense of smell is well developed in newts and salamanders. Air is pumped through the nostrils by the action of the muscles in the throat, and scents are analyzed in the olfactory part of the brain. Salamanders use smell to find and identify food and to communicate with each other. Using chemical cues, male newts and salamanders can identify members of the same species, their age, and their sex, not only when they are living in water but also by detecting traces of identifying chemicals, or pheromones, in the films of water covering the substrates on which they live.

▼ BRAIN AND NERVES OF HEAD
Newt
The brain is long relative to that of mammals and birds. The forebrain processes information concerning smell (chemical sensing); sight is processed in the midbrain; and the hindbrain coordinates movement. The hindbrain is relatively small, perhaps reflecting amphibians' limited range of movements.

cupola

▲ LATERAL LINE NERVE CELLS
Newts have a row of nerve cells, called a lateral line organ, distributed along each flank. The cells are sensitive to changes in water pressure, and enable the newt to detect movements made by other organisms.

hair cells

nerves

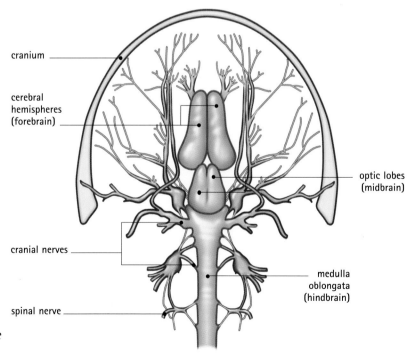

cranium

cerebral hemispheres (forebrain)

cranial nerves

spinal nerve

optic lobes (midbrain)

medulla oblongata (hindbrain)

Circulatory and respiratory systems

COMPARE a newt's heart with that of a *CROCODILE*. The newt's heart, like that of most other amphibians, has a ventricle that is not divided into two parts. A crocodile has a left and a right ventricle.

Adult newts and salamanders absorb oxygen and dispose of carbon dioxide across three surfaces: the lungs, the skin, and the lining of the mouth. In larvae (and neotenic adults), gaseous exchange also takes place in the gills. Most larvae have functional lungs, but in some species these do not develop until the animal begins to metamorphose.

The relative importance of the various ways in which oxygen is taken up by an individual newt or salamander varies according to where it is living at the time, the temperature of the environment, and what stage the animal has reached in its life cycle. In addition, there are variations between species.

Air contains about 30 times more oxygen than an equivalent volume of water, but the precise amount of oxygen dissolved in water varies with temperature: cold water holds more oxygen than warm water. Newts and salamanders living in water have to pass more water over their respiratory surfaces than land-living species do with air, and those living in warm water have to work especially hard to obtain enough oxygen. If they have lungs, they will need to make more trips to the surface to take in air. Species living in very cold water often obtain all, or nearly all, the oxygen they need without coming up for air.

Size is another important variable that affects respiration. Small objects have a greater ratio of surface area to volume than large ones (for a given shape). Small newts and salamanders have a relatively large surface area of skin available for cutaneous respiration, especially if they are long and slender. Large, stout species have a relatively small surface area and need to use their lungs more. Lungless salamanders are able to obtain all their air through cutaneous respiration; and if they are living in water, they never need come to the surface for air.

Newts and salamanders can often be seen flexing muscles in the throat in a rhythmic pumping action. This behavior serves to shunt

▶ CIRCULATORY SYSTEM
Newt
Newts and salamanders that have functioning lungs have a similar heart structure: the left and right atria are separated by an interatrial septum, and the ventricle has no internal subdivision. Most amphibians are unique among air-breathing vertebrates in having no division of the ventricle.

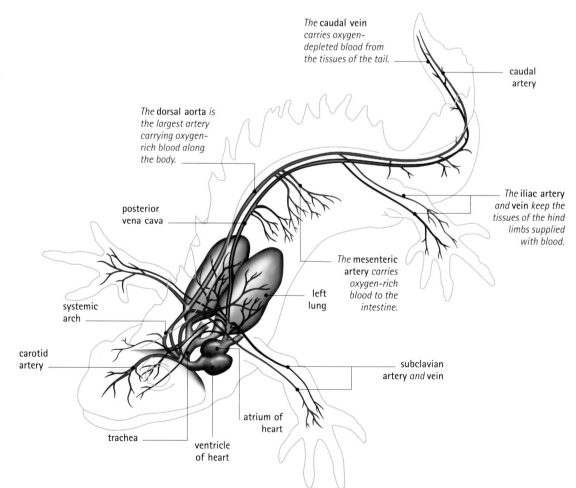

The **caudal vein** carries oxygen-depleted blood from the tissues of the tail.

caudal artery

The **dorsal aorta** is the largest artery carrying oxygen-rich blood along the body.

posterior vena cava

The **iliac artery** and **vein** keep the tissues of the hind limbs supplied with blood.

The **mesenteric artery** carries oxygen-rich blood to the intestine.

left lung

systemic arch

carotid artery

subclavian artery and vein

trachea

atrium of heart

ventricle of heart

Frilly edges

Some salamander species, such as the hellbender, have many folds and frills in their skin. The folds help to increase the surface area available for gas exchange. Hellbenders live in cold, highly oxygenated shallow rivers in the eastern United States; and, although they are large, they obtain most of their oxygen through their skin. The tall crest of the male crested newt serves a similar purpose and enables the newt to remain underwater longer; it is able to complete its courtship without breaking off to return to the surface of the river.

air into and out of their lungs, and some gaseous exchange also takes place across the lining of the animal's mouth, which is served by a rich supply of blood vessels.

Like all amphibians, but unlike birds and mammals (including humans), newts and salamanders are ectotherms: their body operates at much the same temperature as the surroundings rather than producing heat of its own. Amphibians can regulate their body temperature only by moving from warm to cool places and back again. Whereas reptiles often strive to raise their body temperature by basking in sunlight, newts and salamanders prefer a cool body temperature, which they can achieve only by living in cool places. When the weather is hot, newts retreat into crevices and burrows and become largely inactive. Warm weather often produces dry conditions; and because newts and salamanders need a moist skin to allow gaseous exchange to occur, they quickly become too dry (dehydrate) in a dry environment. In addition, a warm body temperature raises the newt's metabolic rate so that it uses more oxygen than it can easily absorb from the environment.

Circulation and blood supply

The circulatory system is strongly linked to respiration, since that is a means of moving oxygen to the parts of the body where it is needed. Newt and salamander hearts have three chambers: two atria receive incoming blood, and a single ventricle pumps blood out of the heart and around the body. Blood is oxygenated in the lungs and flows to the left atrium.

Blood from the head, body, and skin flows into the right atrium. Since the two atria are connected, the blood from different sources may become partially mixed. The blood then passes to the ventricle, where it is pumped back around the body. This system is not very efficient compared with that of mammals and birds, which have a heart with four chambers that keeps oxygen-rich and oxygen-depleted blood separate. However, newts and salamanders have the advantage of being able to absorb oxygen directly through their skin. This feature compensates for their inefficient circulatory system. Indeed, the lungless salamanders make up the most successful salamander family in terms of numbers of species.

▲ *While underwater, crested newts are able to meet their oxygen requirements through cutaneous respiration (across the skin) and buccopharyngeal respiration (through the lining of the mouth).*

In hot water

Aquatic newts and salamanders suffer when the temperature of their environment climbs, because the oxygen-carrying capacity of water decreases as it warms. Such conditions soon cause problems for these amphibians, and the only way they can find relief is to move to a cooler place.

Reproductive system

Newts and salamanders have testes or ovaries, depending on whether they are male or female. Males lack copulatory organs—such as a penis—even though fertilization is internal. Both sexes have a common opening, a cloaca, to the reproductive and digestive systems. Reproduction in the crested newt, and most other species, is seasonal, and takes place in the water. During the summer, males and females feed intensively and produce sperm and eggs, or ova. The newts hibernate and emerge, ready to breed, the following spring. The male crested newt starts to grow a crest during fall and winter; it attains its full glory soon after the male enters the ponds where the newts breed.

Male newts produce a small, conical packet of sperm, or spermatophore. The female has to pick up the spermatophore in her cloaca before she can fertilize her eggs. Depositing a spermatophore is risky because the packet can easily be swept away in a current of water. To increase the chance that a female will find and take up a spermatophore, the male crested newt performs a courtship ritual in which he

▶ MATING
AND LARVA
Red-spotted newt
A pair of newts engage in an underwater courtship dance. After mating, the female lays her eggs, one at a time, attached to aquatic plants. Each newt egg hatches into an aquatic larva, or tadpole (far right). The feathery gills allow the young newt to breathe underwater and so avoid dangerous trips to the surface.

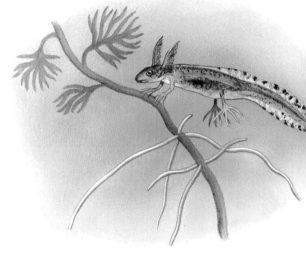

▶ MALE AND FEMALE
REPRODUCTIVE
SYSTEMS
Newt
The male's testes make sperm. In the cloaca, sperm are packaged into spermatophores. The male deposits these in front of the female, who stores them in her cloaca. Eggs, made by the ovaries, are fertilized by the sperm when conditions are favorable.

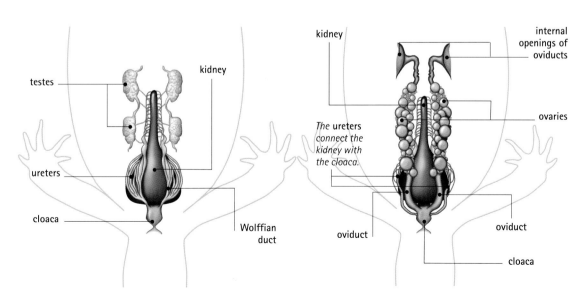

Peter Pan salamanders: larvae that do not grow up

Members of several families of salamanders sometimes exhibit what zoologists call neoteny: individuals remain in the water and retain a larval form throughout their life, although they still become sexually mature. There may be advantages to remaining in the water: the surrounding habitat may be unsuitable because it is too dry, is too cold, or does not contain enough prey animals, for example. Newts living at high altitudes are more likely to be neotenous than populations living elsewhere. There is a neotenous population of crested newts in Sweden, and odd individuals turn up occasionally in other places, although they are rare. There are also several neotenous populations of the Italian crested newt in Montenegro in southeastern Europe.

The most famous neotenous salamanders are the axolotls in Lake Xochimilco, Mexico, which are now very rare in the wild but are widely bred in captivity. In natural conditions they never metamorphose. Several other species of salamanders in the same genus (*Ambystoma*), including the American tiger salamander, also occasionally produce neotenous individuals.

▶ *Mexican axolotls live all their life in water. In the wild they never metamorphose, so they retain their frilly gills— typical of the larvae of other species—after they have become sexually mature. This individual is a laboratory specimen; in the wild, axolotls are dark gray or brown.*

leads the female to the place where he has left it. The male uses a variety of postures and secretes chemicals called pheromones to hold the female's attention. Sometimes other males interfere with the first male's courtship and sneak in at the last minute to deposit their spermatophores on top of the first one.

Once the female newt has picked up the spermatophore, individual sperm are released, and they immediately swim to a chamber in the roof of the cloaca called a spermatheca. There they remain until the eggs are about to be laid. The sperm can remain viable in the spermatheca for up to several weeks; in some salamanders, the spermatophore is taken up by the female in the fall and stored through the winter before being used to fertilize eggs the following spring. Female crested newts, however, usually begin laying eggs shortly after

courtship. As the eggs move down the oviduct, they pass the spermatheca and are fertilized. The crested newt lays about 200 eggs on aquatic vegetation and carefully folds the leaf of an aquatic plant around each one. Egg-laying takes several days. Other salamanders lay their eggs in groups. The time taken for hatching and development depends on the temperature of the water. Under cold conditions, some of the young do not metamorphose until the following year; they live through the winter as well-grown larvae.

CHRIS MATTISON

FURTHER READING AND RESEARCH

Stebbins, R. C., and N. W. Cohen. 1995. *A Natural History of Amphibians*. Princeton University Press: Princeton, NJ.

Newts and Salamanders: www.caudata.org

Octopus

PHYLUM: Mollusca CLASS: Cephalopoda
ORDER: Octopoda

There are about 200 species of octopuses living in marine habitats throughout the world. All octopus have eight arms connected directly to the head, and most are benthic—they live on the sea bottom. The common octopus is widely distributed across the world's oceans, from mild temperate to tropical waters. The octopus lives at depths to about 650 feet (200 m) and spends most of its time on the seabed, but it can also swim.

Anatomy and taxonomy

Scientists categorize all organisms into taxonomic groups based partly on anatomical features. Octopuses, cuttlefish, squid, and nautiloids are cephalopod mollusks. They belong to the class Cephalopoda within the major animal group, or phylum, Mollusca.

● **Animals** Octopuses, like other animals, are multicellular and acquire their food by consuming other organisms. Animals differ from other multicellular life-forms in their ability to move from one place to another (in most cases, using muscles).

● **Mollusks** There are at least 90,000 living species that belong to the phylum Mollusca. The largest groups of mollusks, in terms of number of species, are the class Gastropoda (marine, freshwater, and land-living snails and slugs), with almost 80,000 species; and the class Bivalvia (bivalves, such as clams, mussels, oysters, and scallops), with about 10,000 species. The class Cephalopoda (cephalopods, including the nautiloids, cuttlefish, squid, and octopuses) has just 700 living species, but is the third largest.

Mollusks have a body plan that is not shared by any other animal group. It is based on four regions: a head at the leading edge; a muscular foot on the under (ventral) side for creeping about; a central region called the visceral mass, which contains most of the internal organs; and an upper (dorsal) region that secretes a chalky shell. These body regions are most obvious in gastropod mollusks such as snails and slugs. In bivalves and cephalopods, the body regions can be traced but are highly modified. In aquatic mollusks, a fleshy mantle overhangs the body, creating a space—the mantle cavity—which opens to the environment and into which waste substances from the gut and kidneys are ejected. The mantle cavity also contains gills called ctenidia. All mollusks except bivalves possess a structure

▼ *Octopuses are invertebrates (animals without a backbone), belonging to the group called mollusks, which also includes snails, slugs, and bivalves, such as oysters. The closest relatives of octopuses are squid, cuttlefish, and nautiloids. About 200 species of octopuses live in seas and oceans around the world.*

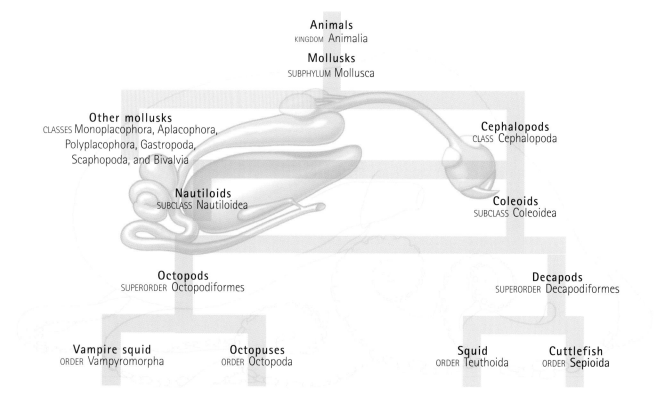

Animals
KINGDOM Animalia

Mollusks
SUBPHYLUM Mollusca

Other mollusks
CLASSES Monoplacophora, Aplacophora, Polyplacophora, Gastropoda, Scaphopoda, and Bivalvia

Cephalopods
CLASS Cephalopoda

Nautiloids
SUBCLASS Nautiloidea

Coleoids
SUBCLASS Coleoidea

Octopods
SUPERORDER Octopodiformes

Decapods
SUPERORDER Decapodiformes

Vampire squid
ORDER Vampyromorpha

Octopuses
ORDER Octopoda

Squid
ORDER Teuthoida

Cuttlefish
ORDER Sepioida

called a radula, which is like a tongue covered with small teeth. Radulae are used for rasping at food to break it down into small particles for swallowing.

● **Cephalopods** The term *cephalopod* comes from the Greek *kephale*, meaning "head," and *podos*, meaning "foot." Cephalopods are predators that swim by jet propulsion. Most have a complex brain and sophisticated eyes. Many cephalopods have lost, through the course of evolution, all or most of the chalky shell found in other mollusks. The mantle cavity faces forward and opens just behind the head. It has a siphon, or funnel, which is part of the foot and through which water is expelled to propel the animal backward. Muscular arms originate from the foot and surround the mouth. All cephalopods have 8 or 10 appendages with suckers (arms or tentacles), except the nautiloids, which have 60 to 90 suckerless tentacles.

● **Nautiloids** The subclass Nautiloidea has only six living species; all are members of the genus *Nautilus*. The nautiloids are slow-moving predators that have retained the outer coiled shell of their ancestors. The shell is divided internally into several chambers, or compartments, separated from one another by walls called septae. The living animal occupies the largest and outermost chamber. A thread of tissue, called the siphuncle, runs from the outer chamber, through all the others, to the center of the shell. Nautiloids can alter their buoyancy by changing the

▲ *An octopus's eight arms surround the bulbous head and are connected by a web of skin.*

proportion of liquid and gas in the shell chambers. In nautiloids, blood circulates through blood spaces (sinuses) rather than through a closed system of blood vessels.

● **Coleoids** The subclass Coleoidea includes all the fast-moving cephalopods: cuttlefish (Sepioida), ram's horn shells (Spirulida), dumpling squid (Sepiolida), true squid (Teuthoida), the vampire squid (Vampyromorpha), and octopuses (Octopoda). In coleoids, the shell is reduced in size and is internalized, or, as in most octopuses, is absent altogether. Blood circulates through a system of closed vessels, and there is one pair of ctenidia.

● **Cuttlefish** Cuttlefish (order Sepioida) have slightly flattened bodies with side fins. They have eight arms and two long tentacles. The long tentacles, which can be withdrawn into pits in the body, bear suckers only on their spoon-shaped tips. The chalky shell, or cuttlefish "bone," is internal and acts as a buoyancy device.

● **True squid** These squid (order Teuthoida) possess cylindrical bodies with side fins. The eight arms and two elongated tentacles bear suckers, which in many species have hooks. The long tentacles are nonretractable. The chalky skeleton is internal and has a penlike shape.

● **Octopuses** Octopuses (order Octopoda) have a squat, rounded body formed of the head and mantle, and eight suckered arms joined to each other and to the head by a web of skin, or interbrachial web. Most octopuses live on the seabed, unlike cuttlefish and squid, which live in the water column.

EXTERNAL ANATOMY The large eyes, horny beak, and suckered arms are clear indications of the octopus's active predatory lifestyle. *See pages 814–817.*

MUSCULAR SYSTEM The octopus walks over the seabed on its arms and swims by jet propulsion. *See page 818.*

NERVOUS SYSTEM The octopus has arguably the most complex brain of any invertebrate and eyes that are similar in general design to those of vertebrates. *See pages 819–820.*

CIRCULATORY AND RESPIRATORY SYSTEMS Unlike other mollusks, octopuses, cuttlefish, and squid have a closed circulatory system and three hearts that deliver blood at high pressure to sustain their active lifestyles. *See page 821.*

DIGESTIVE AND EXCRETORY SYSTEMS The digestive system is simple, for processing the protein-rich diet of a predator. *See page 822.*

REPRODUCTIVE SYSTEM Unlike other mollusks, cephalopods are highly mobile as adults and do not have a larval stage. Eggs hatch into juveniles that are a miniature version of the adult. *See page 823.*

External anatomy

COMPARE the arms of an octopus with the tentacles of a *SEA ANEMONE* and a *JELLYFISH*.

COMPARE the color-changing ability of an octopus with that of a *JACKSON'S CHAMELEON*.

In comparison with most other animals, octopuses have an amazingly variable body shape that changes continually as the animal moves. The upper part of the body is a muscular bag called the mantle, which encloses the animal's vital organs. Around the edge of the mantle lies a chamber, called the mantle cavity, which opens into the surroundings through a siphon, or funnel. When swimming, the octopus jet-propels itself by contracting the mantle and forcing water out through the siphon under high pressure. The jet also flushes out waste products that empty into the mantle cavity from the gut and kidneys.

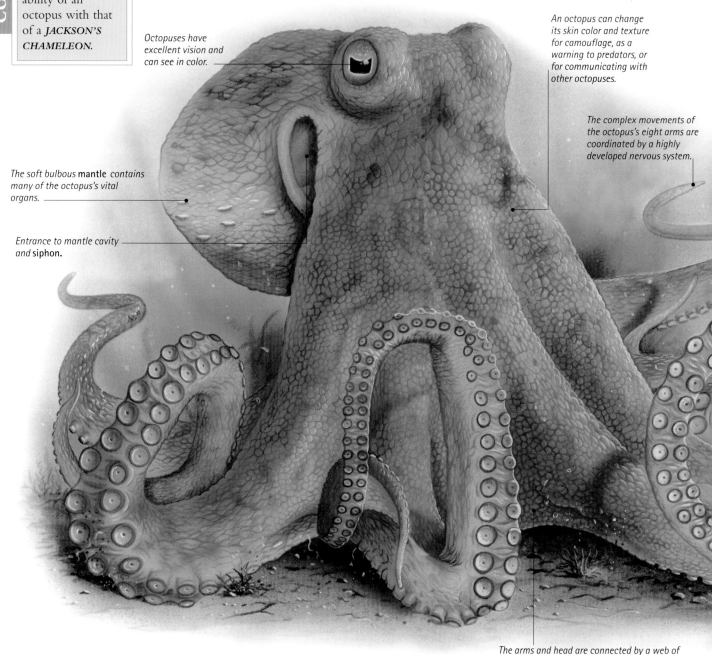

Octopuses have excellent vision and can see in color.

An octopus can change its skin color and texture for camouflage, as a warning to predators, or for communicating with other octopuses.

The complex movements of the octopus's eight arms are coordinated by a highly developed nervous system.

The soft bulbous **mantle** contains many of the octopus's vital organs.

Entrance to mantle cavity and **siphon**.

The arms and head are connected by a web of skin called the **interbrachial web**.

The head is indistinct and continuous with both the mantle on the upper side and the eight arms on the underside. The mouth lies beneath, surrounded by the arms, and is guarded by a vicious-looking hooked beak behind which lies the tonguelike radula with its rasping teeth. The beak bites into soft flesh and cracks hard shells, whereas the radula is a grinding tool that reduces flesh to a soupy pulp.

The common octopus is highly variable in its coloration—typically in the range of gray, yellow, brown, or green—depending on the environment in which the animal lives. For camouflage, octopuses change skin color and pattern to match their background. By

◀ **Common octopus**
Like all other octopuses, the common octopus has eight arms surrounding a fleshy mantle and connected by a web of skin. The underside of each arm has two rows of suckers; and the skin, which has a warty appearance, can abruptly change across a wide spectrum of color and pattern.

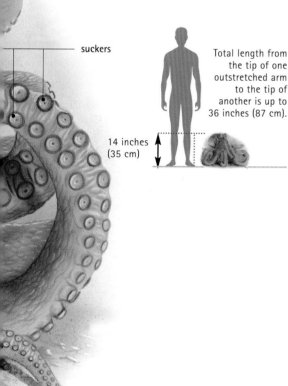

suckers

Total length from the tip of one outstretched arm to the tip of another is up to 36 inches (87 cm).

14 inches (35 cm)

arm

Ammonites

Ammonites belong to a long extinct third subclass of cephalopods, the Ammonoidea. With their spiral shells, ammonites superficially resembled nautiloids. In ammonites, however, the joints between septa and shell are convoluted, and the siphuncle lies against the outer wall of the shell. Ammonites evolved later than nautiloids, but died out at the end of the Cretaceous period—about 65 million years ago—at the same time the dinosaurs became extinct. In their heyday, during the Jurassic period more than 144 million years ago, ammonites were probably the most abundant large predators in the sea. Some grew shells more than 10 feet (3 m) in diameter.

▼ *This fossil ammonite from the Jurassic period has a spiral shell structure similar to that of the nautiloids.*

Buoyancy in nautiloids

Nautiloids can adjust their buoyancy with great speed and precision. The siphuncle of nautiloids adjusts the amount of fluid in the shell chambers. By adding fluid, it reduces the volume of gas in the chambers and makes the shell heavier, or less buoyant, so that the animal sinks. By withdrawing fluid, the siphuncle makes the shell more buoyant, and the animal rises. Using this mechanism, nautiloids living just beyond coral reefs in parts of the South Pacific descend to about 1,150 feet (350 m) by day—well away from shallow-water predators such as dolphins and turtles—and then rise to about 500 feet (150 m) at night to feast on crustaceans and small fish.

▶ ARM

An octopus's arm is covered with two rows of suckers that the animal uses for touching and for holding onto prey. If an octopus loses part of an arm, the missing section will grow back.

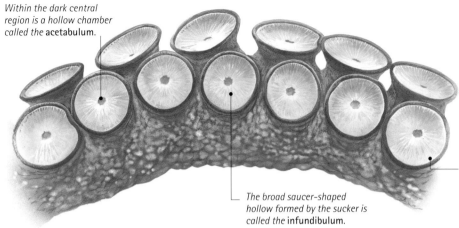

Within the dark central region is a hollow chamber called the **acetabulum.**

The broad saucer-shaped hollow formed by the sucker is called the **infundibulum.**

The **suckers** *contain sensory cells that respond to touch and, in some octopuses, to taste, enabling the animal to detect prey.*

contracting muscles in the skin, they are even able to produce "gooseflesh" and mimic the texture of their background. Through color and pattern, they also communicate aggression, to ward off predators or rivals; or attraction, when courting a mate.

Suckers

Suckers run in two rows on the underside of each arm. Suckers are used to grasp onto prey, but in many shallow-water octopuses, the suckers are also chemosensory—they contain cells that detect chemicals—and the

CLOSE-UP

Changing color

Most octopuses, squid, and cuttlefish that live on shallow seabeds or near the sea surface show astonishing changes in body color and pattern. Cephalopods produce rapid color change by the nervous control of pigment cells in the skin, called chromatophores. Attached to the corners of each chromatophore cell are tiny muscles, which, when they contract or relax, alter the distribution of pigment. When contracted, the muscles pull the cell out into a flat plate, which spreads out the pigment and displays it. When relaxed, the muscles cause the cell to fatten and the pigment to concentrate

into a small, barely visible dot. Different chromatophore cells contain different pigments—black, yellow, orange, red, or blue. The arrangement of different types of chromatophores alongside and above one another, and their degree of expansion or contraction, produces countless permutations of color and pattern. The possibilities are extended further by reflective cells beneath the chromatophores, called iridocytes, that both bend (refract) and bounce back (reflect) light rays.

▶ *Cuttlefish are masters of rapid color change. When traveling across the sea floor, they can instantly change the color of their skin to match their surroundings and thus avoid being seen by predators.*

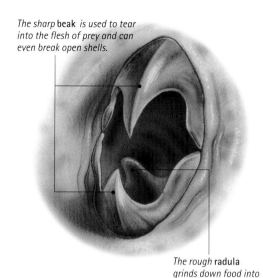

*The sharp **beak** is used to tear into the flesh of prey and can even break open shells.*

*The rough **radula** grinds down food into smaller morsels.*

▲ OCTOPUS BEAK

An octopus's beak is positioned at the entrance to the mouth on the underside of the octopus. The sharp beak is made from a tough substance called chitin, which is also the material that an insect cuticle contains.

animal searches in crevices for food by "blindly tasting" with its arms. The skin of the arms also contains touch-sensitive cells that can detect texture. The cells enable the octopus to identify objects by touch alone. Some squid and cuttlefish have grasping hooks on their suckers, but octopuses do not possess these. In male octopuses, one of the arms tips is modified into a clublike structure called a hectocotylus, which is used to transfer packets of sperm to the female.

The muscular arms can flex and turn in any direction for moving the octopus across the sea bottom, exploring its surroundings, and grasping prey. A muscular web of skin, the interbrachial web, joins the arms together around the mouth, and can surround captured prey to prevent their escape. Some shallow-water octopuses extend their interbrachial web like a sail to drift on water currents near the seabed. Some deep-sea octopuses live in the water column, and their web acts as a parachute to slow their descent when they periodically stop swimming.

IN FOCUS

Ink

Without a shell to protect them, soft-bodied squid, cuttlefish, and octopuses make inviting meals for predators. An octopus's defenses include a biting beak, venom, and a tangle of suckered arms. However, avoiding or escaping an encounter is a safer option than fighting off an attack. Most octopuses conceal themselves in caves or crevices or camouflage themselves when out in the open. However, should concealment or camouflage fail, they can escape under cover of sudden darkness. When threatened, most octopuses produce a dense cloud of brown-black ink, which temporarily hides them from view and disorients the attacker. The ink comes from an ink sac connected to the rectum and is squirted through the mantle cavity. In some species, the ink contains unpleasant substances that have a greasy texture or a foul smell. In the deep ocean, where no sunlight penetrates, producing dark ink would not work as a defense. Instead, some deep-sea octopuses produce ink containing bioluminescent (light-generating) bacteria. The glowing ink cloud temporarily dazzles the attacker while the octopus makes its escape.

▼ *An octopus making a quick getaway from an inquisitive diver ejects a dense cloud of ink to hide itself from view.*

Eyes

Octopus eyes are large and superficially similar to those of vertebrates such as fish. Shallow-water octopuses are relatively nearsighted but can tell objects apart by shape, color, and patterning. They can spot a still or moving object less than 0.2 inch (0.5 cm) across at a distance of more than 3 feet (0.9 m). Information relayed from the eyes to the brain enables the octopus to undergo remarkable changes (for camouflage or communication) in skin color and texture at a speed unrivaled in the animal kingdom.

Muscular system

All living cephalopods, except nautiloids, have dispensed with armor in favor of speed. Squid and cuttlefish have a streamlined body with a reduced chalky internal skeleton. Most octopuses, however, lack any trace of a solid skeleton, and as a result they have an astonishing ability to change shape and squeeze through tiny gaps. For example, a common octopus with a resting body 10 inches (25 cm) across can squeeze through an aperture less than 2 inches (5 cm) wide.

Jet propulsion

Cephalopods use jet propulsion for swimming. The wall of the siphon and the mantle contains bands of circular muscle. When these muscles contract, they squeeze on the mantle cavity and siphon, sending a jet of water out through the siphon opening. The octopus turns the siphon to steer. Longitudinal muscle fibers run at right angles to the circular muscle, and when these contract they cause the mantle cavity and siphon to enlarge, and water flows in around the head to replace the water pumped out.

▼ This Pacific giant octopus is scavenging on a dead dogfish. Muscles in the octopus's suckers enable them to grip the fish tightly.

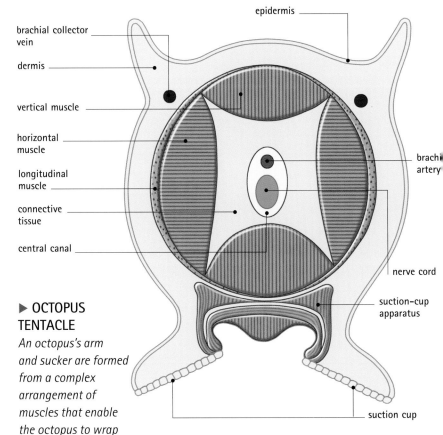

epidermis

brachial collector vein

dermis

vertical muscle

horizontal muscle

longitudinal muscle

connective tissue

central canal

brachial artery

nerve cord

suction-cup apparatus

suction cup

▶ **OCTOPUS TENTACLE**
An octopus's arm and sucker are formed from a complex arrangement of muscles that enable the octopus to wrap around and grip very tightly onto prey.

Cuttlefish and squid have fins that run along the sides of the body. These act rather like wings that the animal uses to stabilize itself as it pushes through the water. They keep the body from twisting and turning but can be angled to make controlled high-speed turns. The fins also flex and ripple, allowing the animal to swim forward or backward slowly, and hover quite precisely, without using the siphon.

Suckers

Octopuses grasp onto prey by using their suckers. Each sucker consists of a fleshy ring surrounding an indentation. When the fleshy ring is applied to an object, circular muscles in the indentation contract, creating an inwardly arched dome that forms a partial vacuum. Water pressure outside the sucker pushes the ring against the object, creating a firm but temporary attachment.

Nervous system

The cephalopod brain is the largest and most complex of any invertebrate. In fact, brain weight relative to body weight in an octopus is greater than that of most fish, amphibians, and reptiles.

The brain processes sensory data, coordinates responses, and in more complex animals—such as most cephalopods and vertebrates—contains "higher centers" concerned with learning and long-term memory. The brain connects to sensory organs—such as touch receptors and eyes—through sensory nerves. The brain also connects to responding organs, or effectors—such as muscles and glands—through motor nerves. Electrical stimuli, in the form of nerve impulses, carry messages from sensory organs to the brain along sensory nerves, and from the brain to effectors along motor nerves.

The cephalopod brain, though complex, is derived from the normal plan of a mollusk nervous system, but with a much greater concentration of nerve cells in the head region, and a vast increase in the number of nerve cells overall. These features coordinate faster and more refined movement and much more complex behavior than in any other mollusk.

The entire nervous system of the sea slug *Aplysia*, a gastropod mollusk, contains about 20,000 nerve cells, whereas the brain of the common octopus contains more than 500 million cells.

The octopus brain is arranged in a dozen distinct regions called lobes. Those lying above the esophagus process sensory information. In particular, the two optic lobes, which receive input from the adjacent eyeballs, are extremely large, showing the importance of vision. Buccal nerves from the buccal lobe at the front of the brain control movement of the radula, and also transmit information on taste.

The brain lobes lying beneath the esophagus coordinate motor responses. Pedal lobes, which are associated with the foot, supply nerves to the siphon, and a single brachial nerve supplies each arm of the octopus.

There are two stellate (star-shaped) nerves that run backward from the brain and connect to two stellate ganglia (clusters of nerve cells). These form the nervous supply to the mantle muscles that control jet propulsion. Two visceral nerves run back to visceral ganglia that connect to the internal organs.

CONNECTIONS

COMPARE the eye of an octopus with the much simpler eye of another mollusk, the **GIANT CLAM**.

COMPARE the brain of an octopus with that of a vertebrate such as a **MANDRILL**. The cephalopod brain is a ring of nervous tissue surrounding the esophagus and is very different in form from that of vertebrates.

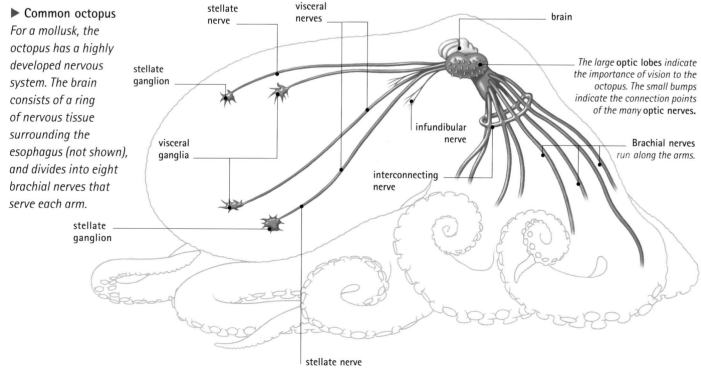

▶ **Common octopus**
For a mollusk, the octopus has a highly developed nervous system. The brain consists of a ring of nervous tissue surrounding the esophagus (not shown), and divides into eight brachial nerves that serve each arm.

stellate nerve

visceral nerves

brain

stellate ganglion

The large optic lobes *indicate the importance of vision to the octopus. The small bumps indicate the connection points of the many* optic nerves.

visceral ganglia

infundibular nerve

stellate ganglion

interconnecting nerve

Brachial nerves *run along the arms.*

stellate nerve

Convergent evolution

Some of the sensory organs in the octopus show remarkable parallels with those in vertebrates. They are examples of convergent evolution—structures with different origins that have evolved to become alike. Convergent evolution occurs when dissimilar organisms respond to similar environmental conditions by evolving comparable structures.

Near the octopus's brain are organs of balance called statocysts, which move relative to the pull of gravity and detect the orientation and acceleration of the head and body. Statocysts resemble structures called otoliths, which are found in the middle ear of vertebrates.

Another example of convergent evolution between the octopus and certain vertebrates is

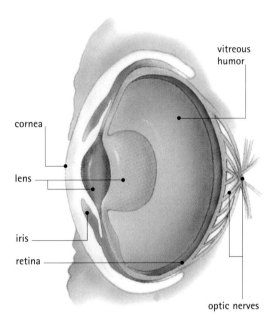

▲ EYE
An octopus's eye is large and resembles that of a vertebrate such as a fish. However, in the octopus the sensory cells of the retina face the front of the eye rather than the back as in vertebrates.

the lateral line system. This consists of rows of receptor cells, containing movable hairs. The rows extend back from the octopus's head and along the arms. As the hairs move, they trigger nerve impulses. The greater the intensity and frequency of vibrations or movement in the water, the greater the frequency of nerve impulses relayed to the brain. This sensory system enables the octopus to detect and respond to disturbances in its surroundings. The system strongly resembles the lateral line system of fish.

Eye structure

An octopus eye looks astonishingly like a fish eye. The biggest structural differences lie in the arrangement of sensory cells and the nervous connection with the brain. In cephalopods, sensory cells face the front of the eye, and the nerve fibers exit from the back of the eye through more than a dozen optic nerves. In vertebrate eyes, the sensory cells face backward (light has to pass through nerve fibers before reaching the light-sensitive regions of sensory cells), and the nerve fibers come together and leave the eye through one optic nerve.

Giant nerves

Cuttlefish and true squid need very fast reflexes to react to the movements of attacking predators or fast-swimming prey. The nervous system contains unusually large nerve cells for high-speed transmission. They cause the mantle muscles to contract simultaneously, resulting in an explosive acceleration from the jet-propulsion system.

In the animal kingdom, two strategies have evolved for high-speed nervous transmission. In fast-moving cephalopods, the nerve cells involved have extra-wide nerve fibers (the elongated part of the cell, also called the axon, that transmits the nerve impulse). Vertebrates' nerve cells, however, have an insulating coating of fat around the nerve fiber. Gaps in the fatty coating cause the nerve impulse to "jump," speeding up the rate of transmission. Fast-conducting nerve fibers in vertebrates are less than 0.0004 inch (0.01 mm) wide. The fast-conducting fibers in a squid are up to 0.04 inch (1 mm) across—they are among the few animal cells large enough to be clearly visible to the unaided human eye.

▼ *The giant nerve fibers of a squid can be seen without a microscope. Thus, squid have been used extensively by scientists seeking to understand the functioning of nervous systems.*

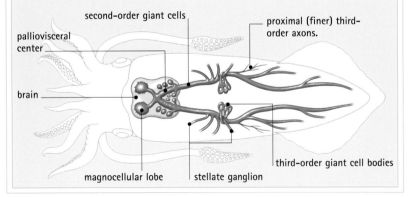

Circulatory and respiratory systems

COMPARE the gills of an octopus with those of a bivalve mollusk such as a *GIANT CLAM*.

COMPARE the closed circulatory system of an octopus with the open circulatory system of another invertebrate such as a *DRAGONFLY*.

As in most other animals, the circulatory system of an octopus delivers food and oxygen to body cells, and takes away waste substances. In noncephalopod mollusks and nautiloids, the circulatory system delivers blood that flows through an open system—in other words, one that is not contained entirely within blood vessels. Such a system carries blood at low pressure, which is enough for slow-moving animals but is insufficient for more active, larger animals. In octopuses, squid, and cuttlefish, a main heart and two accessory hearts pump blood around the body at high pressure through a closed system of blood vessels. The main heart lies in the midline and has three chambers. Two chambers in the heart called auricles receive blood returning from the body circuit and pump it into a single heart chamber, the ventricle, which then pumps blood back around the body. Forcing blood through the gills—or ctenidia—requires high pressure to squeeze the blood through tiny vessels that have a large surface area for exchanging gases. The two subsidiary hearts, one before each gill, carry out this role.

COMPARATIVE ANATOMY

Gills and pigments

In bivalve mollusks such as giant clams, the flow of water over the gills is maintained by wafting hairlike structures called cilia. In cephalopods, mantle muscles maintain the water flow over the gills. The way that the mantle muscles contract ensures that water always flows in the same direction over the gills, and continues whether the octopus is breathing water in or out.

As in other mollusks, the oxygen-carrying pigment in cephalopod blood is copper-containing hemocyanin rather than the iron-containing hemoglobin found in vertebrates. Hemocyanin colors octopus blood blue whereas hemoglobin makes vertebrate blood look red.

The ventilation mechanism coupled with hemocyanin-rich blood and large gills makes gas exchange in the octopus quite efficient.

▶ **Common octopus**
The octopus has a closed circulatory system—the blood is contained within vessels. The system has three hearts. Two of them pump blood through the gills, or ctenidia, where it is oxygenated; and the third, larger heart pumps this oxygenated blood around the body, where the blood gives up its oxygen to cells.

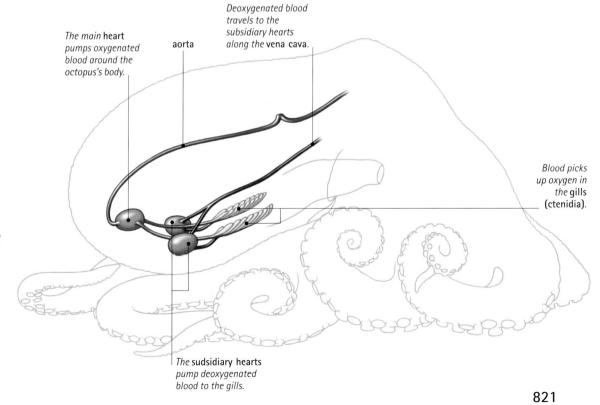

The main **heart** *pumps oxygenated blood around the octopus's body.*

aorta

Deoxygenated blood travels to the subsidiary hearts along the **vena cava.**

Blood picks up oxygen in the **gills** *(ctenidia).*

The **susdiary hearts** *pump deoxygenated blood to the gills.*

Digestive and excretory systems

COMPARE the delivery of venom of an octopus with the methods used by a *GREEN ANACONDA* and a *JELLYFISH*.

COMPARE the crop of an octopus with that of a bird such as an *EAGLE*.

▼ **Common octopus**
Food is first broken into chunks by the beak before it is swallowed. It then passes through the digestive system, where it is further broken down, allowing the absorption of nutrients. Waste products are expelled into the mantle cavity and are blown out into the surrounding water.

Octopuses that live in shallow water tend to hunt on the seabed and swim only occasionally. They eat bivalve and gastropod mollusks, along with crustaceans such as crabs, shrimp, and lobsters. Like cuttlefish and squid, most octopuses hunt by sight and vibration detection at longer range, supplemented by touch, smell, and taste at close range. Most shallow-water octopuses cover their prey with their arms and web, and deliver a venomous bite that contains a nerve agent—a neurotoxin—that immobilizes the victim. The venom is produced along with saliva by two salivary glands at the back of the mouth.

Octopuses open up the shells of bivalve mollusks by clamping on with their suckered arms and then pulling the two halves of the shell apart. Alternatively, they might bore a hole in the shell with the radula and then inject venom through the hole. It is common to find several empty mollusk shells around an octopus lair, some with holes drilled in them.

Enzymes

Some octopuses begin the digestive process while the prey is still enclosed in the web, by releasing saliva containing digestive enzymes from salivary glands at the front of the mouth. The beak of the octopus then tears and slices off chunks of softened flesh, and the radula shreds them before swallowing. Once food enters the mouth, saliva continues the process of chemical

IN FOCUS

Venom

Blue-ringed octopuses, which live in shallow ocean water in parts of the Indo-Pacific, are small—no larger than a human hand—but pack a powerfully poisonous punch. A bite from a blue-ringed octopus can paralyze and kill a person within 15 minutes. The venom is both the octopus's defense and its means of rapidly immobilizing fast-swimming prey, which would otherwise escape. The animal's pattern of blue rings, which become more vivid when it is agitated, serve as a warning to would-be predators.

digestion, which involves gradually breaking down the food into a soluble form. This is later absorbed through the gut and into the blood for delivery to cells around the body.

Swallowed food travels down the esophagus (gullet) and into an expanded region, called the crop, for temporary storage before reaching the stomach. The process of chemical digestion continues in the stomach, which thoroughly mixes the food with digestive enzymes. Smaller particles are diverted into a sac connected to the stomach, called the digestive gland, which produces protein-digesting enzymes. Larger particles remain in the stomach and travel into the first part of the intestine. Digested food is absorbed across the walls of the stomach, digestive gland, and intestine, and from a spiral outgrowth of the stomach called the cecum. Indigestible remains pass through the rest of the intestine and exit through the anus, which empties into the mantle cavity. The two kidneys of coleoid cephalopods deal with soluble wastes. Each kidney receives blood from several veins exiting from each ctenidium (gill). The kidneys filter the blood and dispose of wastes into a renal sac that empties into the mantle cavity.

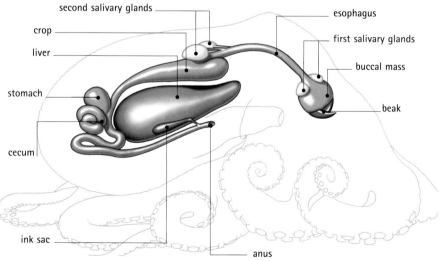

second salivary glands
crop
liver
stomach
cecum
ink sac
esophagus
first salivary glands
buccal mass
beak
anus

Reproductive system

Most octopuses live for only two to five years, reproduce once, and then die. Cuttlefish and octopuses are usually solitary animals, coming together only to mate. Some squid swim in large shoals, but the relationship between individuals is casual and they do not belong to a close-knit social group.

In most cephalopods, females prefer mates that are large: perhaps large size is a sign that the male is a successful hunter, can avoid predators, and has remained alive long enough to grow big. In most species, courtship between male and female is a long-drawn-out affair, with partners holding ritual postures and communicating with each other by complex, shifting color patterns. Individuals can control their color patterns so that the side of the body facing the partner shows a dazzling display, whereas the other side maintains camouflage coloration.

Sperm and eggs

The male octopus manufactures sperm in a single organ, the testis. Millions of sperm leave the testis through the seminal vesicle, which packs the sperm together in an elongated package called a spermatophore, and this is stored in a structure called Needham's sac.

The female produces her eggs in a single organ, the ovary, in her midline. Her released eggs pass through the oviducal gland, which coats them in a protective membrane. Other glands may add further coatings to the eggs before they travel down two oviducts to reach the mantle cavity.

Once mating is under way, the male uses his modified arm, the hectocotylus, to pick up the spermatophore from Needham's organ and, in the case of the common octopus, transfers it to one of the female's oviducts. In some species of octopuses, the tip of the hectocotylus, clutching its sperm packet, breaks off inside the female. The spermatophore becomes primed when it is removed from Needham's sac, and the mass of sperm spring to the outside, together with a cement body, which attaches itself inside the female octopus. The spermatophore gradually disintegrates over a day or two, releasing sperm, which fertilize the female's eggs.

Egg care

In the common octopus, the male leaves the female after mating, and she retreats to a hollow on the seabed where she lays her eggs —more than 100,000—in clusters that she attaches to rocks. She guards the eggs and gently jets water over them to keep them well supplied with oxygen and to clean away waste. The female never leaves the eggs, and as a consequence she starves herself while the eggs are developing. When the eggs hatch and the new generation swims away, the emaciated mother dies.

TREVOR DAY

▲ MATING
After an elaborate courtship, the octopuses mate. The male inserts his hectocotylus into the female's mantle and delivers a packet of sperm, which the female uses to fertilize her eggs.

▼ HECTOCOTYLUS
In some species of octopuses, the tip of the hectocotylus breaks off in the female during copulation.

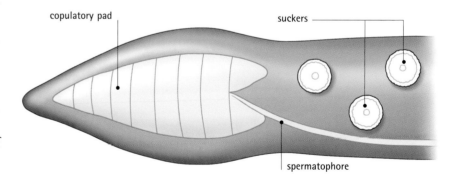

copulatory pad

suckers

spermatophore

FURTHER READING AND RESEARCH

Brusca, R. C., and G. J. Brusca. *Invertebrates* (2nd edition). 2003. Sinauer Associates: Sunderland, MA.

Cephalopod database: www.cephbase.utmb.edu

Orchid

ORDER: Orchidales FAMILY: Orchidaceae
GENERA: *Orchis* and 400–800 others

Orchids are prized for their flowers. Their exotic shapes and colors have encouraged collection and cultivation on a global scale. They are a multimillion-dollar industry—in 2004, more than 17 million orchids were sold in the United States. Many orchids have an unusual lifestyle: about half of all orchid species grow high in trees, rather than in soil. Orchid flowers are extraordinarily varied, but they have a characteristic structure that is very different from other plants and defines the family. To attract pollinators, orchids have evolved some of the most advanced and diverse pollination methods.

Anatomy and taxonomy

All life-forms are classified in groups of relatively closely related species. The classification is based mainly on shared anatomical features, which usually (but not always) indicate that members of a group have the same ancestry. Thus classification shows how life-forms are related to each other.

● **Plants** The first land plants to evolve were spore plants (Pteridophyta), including ferns, liverworts, club mosses, and horsetails. These plants reproduce using spores rather than seeds. Today, the seed plants (Sporophyta) are the dominant type of plants in most habitats. There are two major groups of seed plants: gymnosperms, which have naked seeds, include pine trees and other conifers; and angiosperms (with enclosed seeds) are the flowering plants. Angiosperms' ovules are contained within an ovary, which ripens into a fruit containing the seeds. They are divided into monocotyledons, usually called monocots; and dicotyledons, or dicots. Dicots have two cotyledons, or seed leaves (the first leaves of the plant). Dicots include most trees and many familiar flowers.

● **Monocots** Monocots, which are also called Liliopsida, include palms, grasses, lilies, and orchids. On germinating, most monocot seedlings have a single cotyledon, in contrast to the paired seed leaves of dicots. Although orchids are monocots, most orchids have no cotyledons at all. However, they do share other characteristics of monocots. Orchid leaves are usually long, with parallel leaf

▼ *Orchids form an enormous family of plants, containing many thousands of species. Although they are classified as monocots, most orchids have ovules that lack seed leaves (cotyledons).*

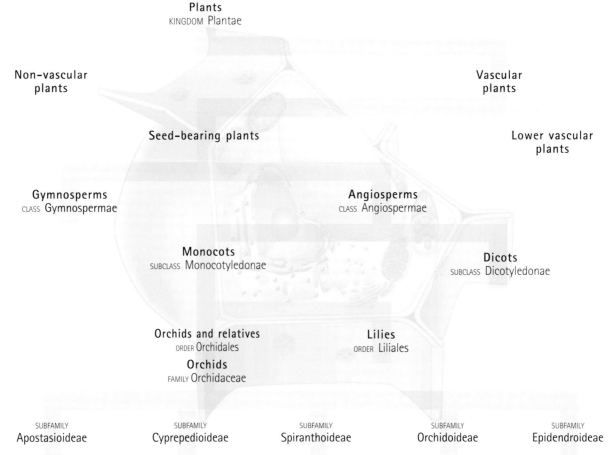

Plants
KINGDOM Plantae

Non-vascular plants

Vascular plants

Seed-bearing plants

Lower vascular plants

Gymnosperms
CLASS Gymnospermae

Angiosperms
CLASS Angiospermae

Monocots
SUBCLASS Monocotyledonae

Dicots
SUBCLASS Dicotyledonae

Orchids and relatives
ORDER Orchidales

Lilies
ORDER Liliales

Orchids
FAMILY Orchidaceae

SUBFAMILY
Apostasioideae

SUBFAMILY
Cyprepedioideae

SUBFAMILY
Spiranthoideae

SUBFAMILY
Orchidoideae

SUBFAMILY
Epidendroideae

veins. The bundles of vessels that transport water and sugar solution through the plant (vascular bundles) are scattered or grouped in two or more rings, rather than in a single ring as in dicots. Also, there is no vascular cambium (one of the important tissues in dicot growth).

Different classifications divide the monocots into various groups. One commonly used system breaks monocots into five subclasses: Alismatidae, including many pondweeds; Arecidae, which includes palms; Commelinidae, including rushes, sedges, and grasses; Zingiberidae (bromeliads and gingers); and Liliidae (orchids and lilies).

• **Liliidae** This group is divided into two orders, Liliales and Orchidales. Plants in the order Liliales (irises, lilies, and onions) have seeds of ordinary size and structure, usually with endosperm (food-storage tissue) and a well-developed embryo (the beginning of new plant). The flowers of these plants often have a radial arrangement with more than one line of symmetry.

• **Orchidales** Plants in the order Orchidales produce vast numbers of minute seeds, with no endosperm and an undeveloped embryo. The plants acquire some food by a fungal partner (they are mycotrophic) and sometimes have no chlorophyll. The flowers are symmetrical but have only one line of symmetry.

• **Orchids** The orchids form the second largest family of plants, Orchidaceae, which has between 15,000 and 35,000 species. Orchids live on every continent except Antarctica, and examples are found in almost every country, except for a few isolated islands. Orchids grow in almost every terrestrial environment, from dry deserts to hot, humid rain forests, soggy bogs, and cold tundra. Around half of all orchids are epiphytes (from the Greek *epi*, meaning "on", and *phytos*, meaning "plant"), living on trees and other plants. Epiphytic orchids have aerial roots

▲ *Orchids are famous for their beautiful flowers, such as this* Phalaenopsis *species. Although very varied, orchid flowers have the characteristic single line of symmetry.*

that supply the orchid with water and minerals without reaching down to the soil.

There are five subfamilies of orchids: Apostasioideae, Cyprepedioideae, Spiranthoideae, Orchidoideae, and Epidendroideae. The subfamily Apostasioideae has only 20 species. Plants in Cyprepedioideae have flowers with a slipper-shaped lip petal, or labellum, and include the temperate slipper orchids and the tropical genus *Paphiopedilum*. The Orchidoideae usually survive adverse periods as dormant tuberoids (fleshy storage roots). In the genus *Orchis*, the tuberoids occur in pairs and are said to resemble testicles. Indeed, the name orchid comes from *orchis*, the ancient Greek word for testicle.

Epidendroideae orchids are usually epiphytes, often with fleshy leaves. The subfamily is divided into 16 groups, with many familiar ornamental orchids, including *Cymbidium*, *Cattleya*, *Dendrobium*, and *Phalaenopsis*. The tribe (group of genera) Arethuseae contains the four species of grass pinks, which live in eastern North America. Unlike most Epidendroideae species, the grass pink orchids are terrestrial, not epiphytic. They grow from an underground storage organ called a corm and have one or two grasslike leaves.

EXTERNAL ANATOMY Orchid plants are very variable in size, flower structure, and color, and way of living. Some are epiphytic, living high in trees and with their roots attached to the trees' bark. *See pages 826–828.*

INTERNAL ANATOMY Orchids have an internal structure typical of monocot plants. The vascular bundles are irregularly spaced throughout the stems, and because there is no secondary thickening, orchids do not grow large and woody. *See pages 829–831.*

REPRODUCTION Orchids have distinctive flowers, with the sexual organs fused into a structure called the column. Orchid seeds are tiny, and each seed pod produces millions of seeds. *See pages 832–833.*

FEATURED SYSTEMS

External anatomy

COMPARE an orchid with *MARSH GRASS*. Both plants are monocots, so they have long leaves with parallel veins. Grasses, however, have tiny flowers with no petals, whereas orchid flowers are almost always bright and showy.

Orchids display an amazing range of shapes, sizes, and growth forms. They vary in size from *Platystele jungermannioides*, which is less than 0.25 inch (6.4 mm) tall, to the massive species in the genus *Grammatophyllum* and the climbing plants of the genus *Vanilla*. The largest orchid is probably a species of *Vanilla*, as the climbing stems can reach many feet up a tree or cliff. *Grammatophyllum* species are the bulkiest orchids, with thick stems up to 16 feet (5 m) long. There are no woody orchids, as monocots have no secondary thickening and so cannot form layers of wood.

Monocot roots

Orchid roots serve the same function as those of other plants: they collect water and mineral nutrients and anchor the plant in its substrate.

IN FOCUS

Roots as storage organs

Some species of orchids store reserves of carbohydrates and other nutrients in specialized roots. In some, the whole root is very fleshy. These fleshy storage roots are called tuberoids to distinguish them from true tubers, which by definition are stems. In *Cleistes* species, some roots or root parts are thick and others much thinner. The thickened regions are called nodules.

Many orchids in the subfamily Orchidoideae have structures called root-stem tuberoids. These are storage roots, but the lower portion has a sheath of root structure around a bud at the tip. This structure survives during the dormant season, after which the bud produces new shoots.

Like other monocots, orchids have no taproot (the single, carrotlike main root). Instead, they produce a tangle of secondary roots. These vary in thickness but are never thin and fibrous like those of grasses.

Orchid roots have a variety of functions. Some orchids use their roots to store carbohydrate reserves. Roots may also function as "trash baskets," storing waste products. Some tropical epiphytic orchids (those that grow on trees), including *Ansellia*, *Cyrtopodium*, and *Grammatophyllum* species, have two types of roots. As well as normal roots, they have a mass of upward-pointing roots that catch dead leaves and other debris falling through the forest canopy. As the debris rots, it forms the orchid's own compost heap, and the roots absorb the nutrients released.

Some orchids have no roots at all. *Cheirostylis* has a fleshy rhizome (underground stem) that bears no roots as such, only root ridges. These

◄ *These dramatically patterned flowers belong to a* Grammatophylum *orchid. The eye-catching coloration helps attract pollinators to the flowers.*

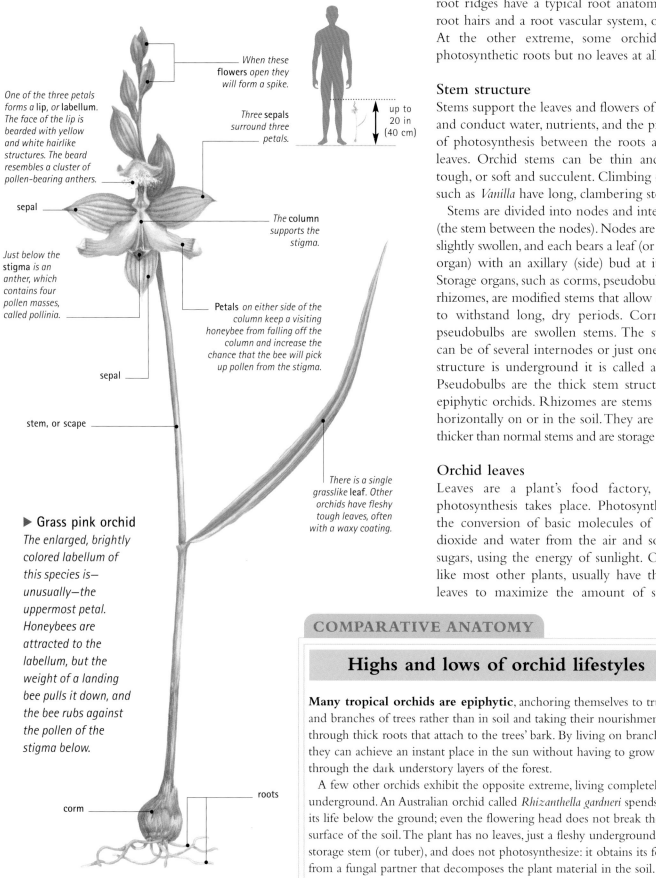

One of the three petals forms a lip, or **labellum**. The face of the lip is bearded with yellow and white hairlike structures. The beard resembles a cluster of pollen-bearing anthers.

sepal

Just below the **stigma** is an anther, which contains four pollen masses, called pollinia.

sepal

stem, or scape

▶ **Grass pink orchid**
The enlarged, brightly colored labellum of this species is—unusually—the uppermost petal. Honeybees are attracted to the labellum, but the weight of a landing bee pulls it down, and the bee rubs against the pollen of the stigma below.

corm

When these **flowers** *open they will form a spike.*

Three **sepals** surround three petals.

up to 20 in (40 cm)

The **column** supports the stigma.

Petals on either side of the column keep a visiting honeybee from falling off the column and increase the chance that the bee will pick up pollen from the stigma.

There is a single grasslike **leaf**. Other orchids have fleshy tough leaves, often with a waxy coating.

roots

root ridges have a typical root anatomy, with root hairs and a root vascular system, or stele. At the other extreme, some orchids have photosynthetic roots but no leaves at all.

Stem structure

Stems support the leaves and flowers of a plant and conduct water, nutrients, and the products of photosynthesis between the roots and the leaves. Orchid stems can be thin and wiry, tough, or soft and succulent. Climbing orchids such as *Vanilla* have long, clambering stems.

Stems are divided into nodes and internodes (the stem between the nodes). Nodes are usually slightly swollen, and each bears a leaf (or leaflike organ) with an axillary (side) bud at its base. Storage organs, such as corms, pseudobulbs, and rhizomes, are modified stems that allow orchids to withstand long, dry periods. Corms and pseudobulbs are swollen stems. The swelling can be of several internodes or just one. If the structure is underground it is called a corm. Pseudobulbs are the thick stem structures of epiphytic orchids. Rhizomes are stems that lie horizontally on or in the soil. They are usually thicker than normal stems and are storage organs.

Orchid leaves

Leaves are a plant's food factory, where photosynthesis takes place. Photosynthesis is the conversion of basic molecules of carbon dioxide and water from the air and soil into sugars, using the energy of sunlight. Orchids, like most other plants, usually have thin, flat leaves to maximize the amount of sunlight

COMPARATIVE ANATOMY

Highs and lows of orchid lifestyles

Many tropical orchids are epiphytic, anchoring themselves to trunks and branches of trees rather than in soil and taking their nourishment through thick roots that attach to the trees' bark. By living on branches, they can achieve an instant place in the sun without having to grow up through the dark understory layers of the forest.

A few other orchids exhibit the opposite extreme, living completely underground. An Australian orchid called *Rhizanthella gardneri* spends all its life below the ground; even the flowering head does not break the surface of the soil. The plant has no leaves, just a fleshy underground storage stem (or tuber), and does not photosynthesize: it obtains its food from a fungal partner that decomposes the plant material in the soil.

▶ Rhizanthella gardneri
This species of Australian orchid is rarely seen, since most of the plant lives below the ground surface. Even the flower cluster remains mostly below the surface.

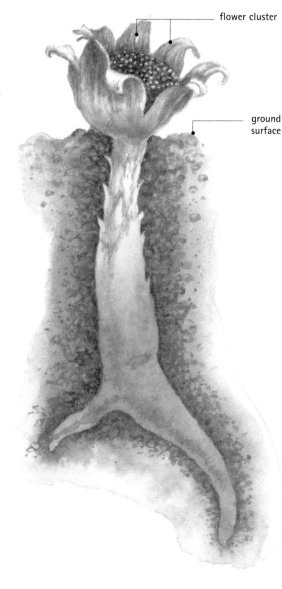

flower cluster

ground surface

▼ RHIZOME
These modified stems lie along or within the ground and are often used by plants to store nutrients.

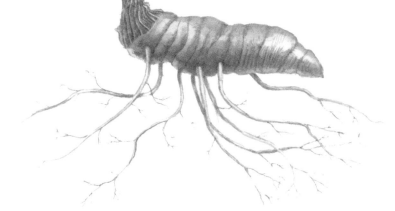

they can capture. The leaves are generally long and narrow, and they are almost always simple in shape with edges that are straight, not serrated or toothed.

Some orchids have a different leaf shape. In *Thelymitra spiralis*, the leaf has a long, twisted tip. Leaves of *Dendrobium cucumerinum* are fleshy, like cucumbers. Other orchids have lost their leaves almost altogether: the leaves are just tiny scales. In *Bulbophyllum globuliforme*, there are no leaves, and both photosynthesis and storage are carried out by pseudobulbs.

Flowers

Orchid flowers present an amazing variety of shapes, forms, colors, and patterns. The basic structure, however, is always the same: on the outside are three sepals, and inside these are three petals. Orchid petals tend to be large and brightly colored. The middle—usually also the lowest—petal is known as the lip, or labellum. This petal is often larger than the other petals, and has a different structure and bright color. The sepal and petal structure produce the flower's single line of symmetry.

The sexual parts of the flower, which in most flowers consist of separate structures for the male stamens and female styles, are merged in orchids into a single structure, called the column, in the center of the flower.

Many orchids have a single flower on each stem; others have flowers borne in a group called an inflorescence, usually in an elongated form called a spike. This arrangement varies, however. The flower head of the underground orchid *Rhizanthella gardneri* is a cluster of 150 or so tiny flowers.

▶ CORM
Like rhizomes, corms are modified stems. Their thickened, bulbous shape provides a convenient nutrient store for the plant.

Internal anatomy

Like all other plants, orchids need to obtain their energy and nutrients from the environment. Orchids' internal anatomy has evolved systems and structures to allow different species to live and reproduce under a wide variety of conditions.

Stems and leaves

Orchid stems have a similar internal anatomy to that of other monocots. Vascular bundles are scattered, in contrast to the ring pattern in dicot stems. The vascular bundles are embedded in parenchyma (unspecialized storage tissue) and contain xylem vessels, phloem vessels, and associated cells. Xylem consists of large, empty tubular cells that conduct water from the roots to the aerial parts of the plant. Phloem cells are long, thin, and living; they transport the products of photosynthesis.

Leaves have an outer layer of cells, called the epidermis, which is covered with a waxy cuticle, or coating. Small pores, called stomata, on the underside of the leaf control the movement of gases into and out of the leaf. The vascular bundles, containing phloem and xylem tissue, are arranged in regularly spaced veins running in parallel lines along the length of the whole leaf. The bulk of a leaf consists of parenchyma cells, which are packed with chloroplasts. Chloroplasts are organelles

COMPARE the thin, fibrous roots of *MARSH GRASS* with the relatively fleshy roots of orchids. Orchids depend on fungi associated with the root to help them search a large area of soil for food and water.

CONNECTIONS

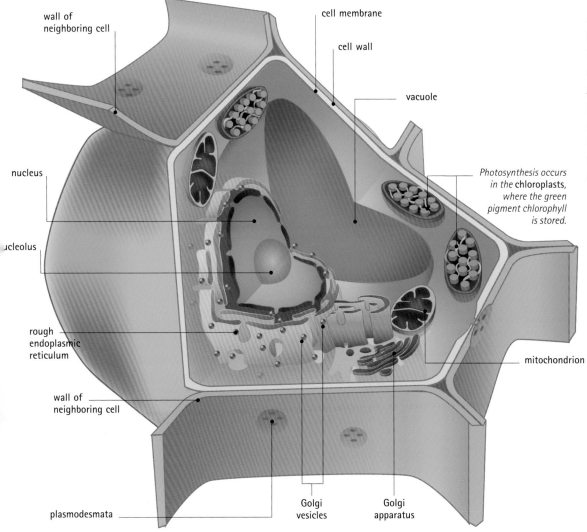

wall of neighboring cell

cell membrane

cell wall

vacuole

nucleus

nucleolus

Photosynthesis occurs in the chloroplasts, where the green pigment chlorophyll is stored.

rough endoplasmic reticulum

wall of neighboring cell

mitochondrion

plasmodesmata

Golgi vesicles

Golgi apparatus

◄ **INSIDE A PLANT CELL**

The cell membrane of a plant cell is surrounded by a cell wall. Inside the cell are all the organelles including the nucleus, chloroplasts, and mitochondria.

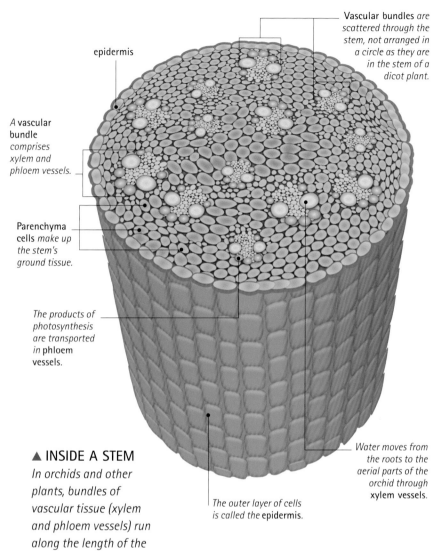

epidermis

Vascular bundles *are scattered through the stem, not arranged in a circle as they are in the stem of a dicot plant.*

A vascular bundle comprises xylem and phloem vessels.

Parenchyma cells *make up the stem's ground tissue.*

The products of photosynthesis are transported in phloem vessels.

Water moves from the roots to the aerial parts of the orchid through xylem vessels.

The outer layer of cells is called the epidermis.

▲ INSIDE A STEM

In orchids and other plants, bundles of vascular tissue (xylem and phloem vessels) run along the length of the stem to transport water and nutrients.

▶ INSIDE A LEAF

This cross section through a leaf of a plant shows the typical arrangement of tissues, from the epidermal cells on the outside to the parenchyma cells in the middle of the leaf and the xylem and phloem cells in the central vein.

(mini-organs inside cells that perform specific functions) containing the green pigment chlorophyll, and are where photosynthesis takes place. Plant cells also contain other organelles including plastids, mitochondria, and a nucleus. Each of these organelles is enclosed by a membrane.

CAM metabolism

In order to photosynthesize, plants need both water and the gas carbon dioxide (CO_2). Epiphytic orchids have a problem: with aerial roots, the supply of water may not be continuous. To obtain CO_2, they need to open their stomata (pores) to the air, but by doing so they lose water vapor. Epiphytic orchids (and many other plant families) have solved this problem by using a chemistry trick. This system was first seen in crassula plants, such as stonecrops and houseleeks, so it is called crassulacean acid metabolism (CAM).

Normally, carbon dioxide is absorbed and converted from a gas to a usable carbon-containing compound only while there is sunlight. Using the CAM pathway, however, plants can absorb CO_2 during the night and turn it into a soluble chemical called malic acid. This allows them to open their stomata only during the night, when it is cooler and water loss is therefore lower. As the sun rises, the plants can close their stomata, having stored enough CO_2 for the day's photosynthesis.

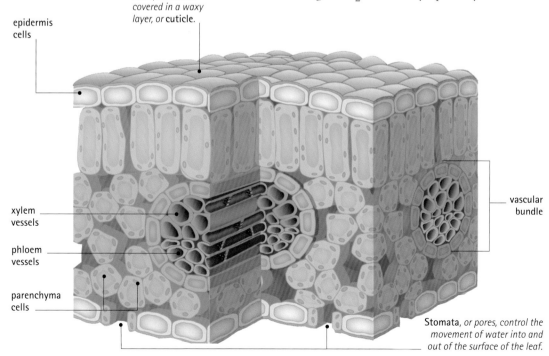

The epidermis is covered in a waxy layer, or cuticle.

epidermis cells

xylem vessels

phloem vessels

parenchyma cells

vascular bundle

Stomata, *or pores, control the movement of water into and out of the surface of the leaf.*

Aerial roots

Most plant roots grow in soil, but those of epiphytic orchids, which live on trees and other plants, never reach the soil. Instead, they are firmly attached to the branches of trees.

Aerial roots of most epiphytic orchid species absorb water and nutrients from the humid air or the surrounding leaf debris. They are covered with a spongy, whitish sheath called the velamen. This consists of an outer layer of cells with partially thickened cell walls, which lose their living contents as the root matures. Because the cells are empty, they act as sponges, holding water between bouts of rain.

▶ *Epiphytic orchids often live high up on trees, far from the soil. Their extensive aerial roots absorb water like sponges, allowing the leaves to produce food by photosynthesis.*

Root structure

Terrestrial orchid roots have root hairs, which are extensions of single cells that reach out into the soil or other substrate. Together, they form a large surface area for absorbing water and mineral nutrients. Epidermal cells form the outer, protective layer of the root. In most orchids, the epidermis consists of one or more layers of cells called the velamen. This layer is especially obvious on the aerial roots of epiphytic orchids. The bulk of the root tissue is made up of cortex—parenchyma cells where the products of photosynthesis are stored. Roots also contains vascular tissue consisting of xylem and phloem vessels, which form a relatively broad ring near the center of the root, called the vascular stele. The center of the ring is filled with pith cells. The endodermis is a cylinder one cell thick that forms a boundary between the cortex and the stele. Water uptake is controlled at the endodermis. Some orchids have enlisted the help of fungal partners, which form a close association with the root called a mycorrhiza (literally, "fungus root").

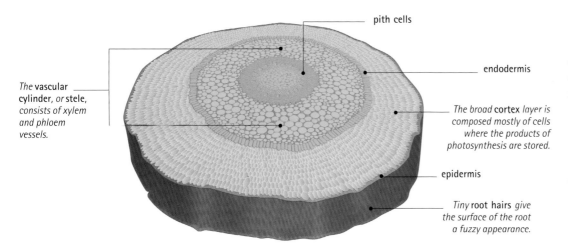

pith cells

endodermis

The **vascular cylinder,** *or* stele, *consists of xylem and phloem vessels.*

The broad **cortex** *layer is composed mostly of cells where the products of photosynthesis are stored.*

epidermis

Tiny **root hairs** *give the surface of the root a fuzzy appearance.*

◀ **INSIDE A ROOT**
A single-cell layer called the endodermis surrounds the stele, which comprises vascular tissue around the central pith. Outside the endodermis is the cortex, which makes up the bulk of the root.

Reproduction

COMPARE orchid seeds with **FERN** spores. Both are incredibly tiny and dustlike, and so are easily dispersed by the wind. Whereas orchid seeds germinate to produce a protocorm, spores germinate to form a structure (gametophyte) that is a separate phase of the life cycle.

Orchids reproduce primarily by sexual reproduction. They display some of the most diverse pollination methods in the plant kingdom. Coevolution (in which two kinds of organisms each influence the evolution of the other) has promoted diversity of both orchid and insect species. As insects have diversified, orchid species have evolved that have close links with their pollinators' behavior, size, and shape.

Asexual reproduction, where it occurs, is usually by tubers and root buds. For example, the bog orchid *Malaxis paludosa* has small projections called bulbils that grow at the tips of leaves and break off to form new plants.

Flower structure

Orchid flowers come in a huge range of shapes, sizes, and colors. However, they all have the same basic structure.

The flower is attached to the main stem by a flower stalk, called a pedicel. The ovary, which becomes the seed capsule, is often very small in orchids, sometimes discernible only as a slight swelling of the pedicel. It contains thousands of ovules that, if fertilized, become seeds.

Orchid petals are actually two different types of structures: three true petals and three sepals. In apples, potatoes, and many other plants, the sepals are the green bracts that protect the flower bud. Orchids and many other monocots have modified sepals that are large and brightly colored, like the true petals. A characteristic of orchids is that one petal, the labellum, is different from the other two. It is usually larger and more complex, and is often of a color contrasting with the rest of the flower. In *Calopogon* and some other species, the labellum, or part of it, is hinged and movable.

The style (the stalk bearing the stigma—the receptive female surface) and stamen filaments (supporting the pollen-bearing anthers) are always merged to some degree. In most species,

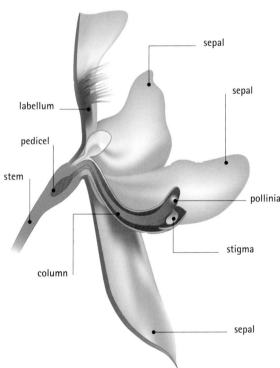

▲ **Grass pink orchid**
This flower has the characteristic structures found in orchid flowers, including colored, petal-like sepals and a style and stamen combined into a single structure, the column. The enlarged (and movable) upper petal, or labellum, of this species is, however, very unusual.

COMPARATIVE ANATOMY

Fooling insects

Many orchids deceive their pollinators by advertising nonexistent rewards. Bee orchids (genus *Ophrys*) and certain tropical orchids have flowers that mimic bees or wasps. Their scents stimulate male bees to attempt sexual union (copulation) with them. During this pseudo-copulation, pollen is transferred to the stigma of the plants from bees. *Corybas* orchids mimic the smell of mushrooms. These attract flies that normally lay their eggs on mushrooms. Similarly, *Bulbophyllum* flowers smell of carrion (rotting flesh) and attract carrion flies. The scents and patterns on the flowers of slipper orchids (*Cypripedium* species) fool insects into exploring them. The flower labellum forms a pouch, or slipper, that acts as a pitfall trap. Once an insect is inside the trap, the only easy route out is through a tight exit. On leaving, the insect usually brushes against first the stigma, then the anther, where it picks up pollen. When the bee visits the next flower, it will leave some pollen on the stigma.

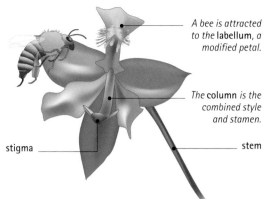

A bee is attracted to the labellum, a modified petal.

The column is the combined style and stamen.

stigma

stem

▲ POLLINATION
Grass pink orchid
Bees are attracted to yellow structures on the hinged labellum of the flower. When a bee lands on the labellum, its weight pulls the labellum downward, forcing the bee against the flower column. There, a packet of pollen, or pollinium, attaches to the bee, which takes it to another flower.

the union is so complete that they are indistinguishable. The combined structure is called the column. The size and shape of the column and the arrangement of the structures on it vary widely between species.

Almost all orchids have pollen grains that are grouped in a waxy mass (pollinium; plural, pollinia), unlike the loose, dry pollen grains of many other plants. The number of pollinia varies among orchid species. This aggregation of pollen is related to the huge number of ovules in the ovary that need to be fertilized. Orchid flowers may also have nectaries (which secrete nectar) and osmophores (scent glands).

Pollination
Most orchids are pollinated by insects, such as bees, butterflies, and moths. An insect in search of nectar or another reward enters the flower, where it comes into contact with the viscidum, a sticky disk connected by a stalk to the pollinia. The viscidum sticks to the insect, taking with it pollinia, which the insect carries to another flower, producing cross-pollination. Pollinia can be attached to the insect on its proboscis, eyes, head, back, or legs, depending on how it enters, where it sits on the flower, or how it leaves.

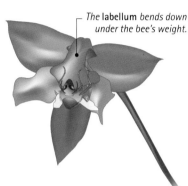

The labellum *bends down under the bee's weight.*

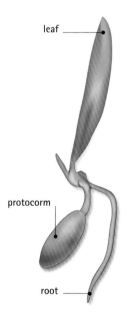

leaf

protocorm

root

▲ SEEDLING
After germination, the orchid embryo swells into a mass of cells called a protocorm. Root hairs and, later, a leaf sprout from the protocorm, forming a seedling.

Orchids and euglossine bees

Euglossine, or orchid, bees pollinate at least 3,000 species of tropical American orchids. Often, only the male bee visits the flower, where it gathers perfume droplets from the surface of the flower. The bee stores the liquid in its hollow hind legs and uses the fragrances as sex chemicals to attract females. *Gongora* and *Coryanthes* orchids are pollinated in this way. Each species produces a perfume specific to a single bee species.

This basic pattern of insect pollination has been modified by orchids into an incredible diversity of elaborate, often deceptive, forms. Orchids use traps; mimic the food, shelter, or mates of insects; or offer sex chemicals that males collect to attract females.

Where a food reward is offered to attract pollinators, the food is principally nectar. However, about one-third of orchids do not fulfill their "promise" of food but deceive the insects. For example, grass pink orchids have yellow hairs on the lip that look like yellow pollen-bearing anthers. These attract bees, which try to collect the pollen for food.

Fertilization occurs several weeks to months after pollination since pollination triggers the maturation of the ovules. Orchid seeds develop within a fruit or seed capsule. The seeds are tiny, like specks of dust. Each pod can produce 2 or 3 million seeds. The tiny seeds are wind-dispersed in all but a few genera. *Apostasia, Selenipedium,* and *Vanilla* have small but relatively heavy seeds that are borne in a somewhat pulpy pods. Vanilla pods are also called vanilla beans, and its scent—which we use as a flavor—attracts seed dispersers.

Because orchid seeds are so tiny, there is no room within them for stored food reserves, or endosperm. For germination, an orchid seed has to form a relationship with a fungal partner that can supply it with nutrients from decaying plant material in soil.

ERICA BOWER

FURTHER READING AND RESEARCH
Heywood, V. H. 2006. *Flowering Plants of the World.* Firefly: Toronto.
Van der Cingel, N. A. 2001. *An Atlas of Orchid Pollination: America, Africa, Asia and Australia.* A. A. Balkema: Netherlands.

Ostrich

ORDER: Struthioniformes
FAMILY: Struthionidae GENUS: *Struthio*

The ostrich is the largest bird in the world, growing up to almost 9 feet (3 m) tall and weighing 300 pounds (135 kg). It is well adapted for life in hot, arid environments such as savanna, dry plains, and semidesert. It lives only in Africa but is bred in captivity elsewhere.

Anatomy and taxonomy
Biologists place organisms into groups primarily according to their anatomy, although studies of DNA are an increasingly important tool in determining relationships.

● **Animals** All animals are multicellular and rely on other organisms for food. They differ from other multicellular life-forms in their ability to move around (generally using muscles) and in their rapid responses to stimuli.

● **Birds** Birds are bipedal (two-legged), warm-blooded vertebrates that lay eggs. They are characterized primarily by their feathers, wings, and lightweight hollow bones. There are almost 10,000 known species, ranging in size from tiny humming-birds to the giant ostrich.

● **Paleognathae** The subclass Paleognathae is made up of several groups (orders) of flightless birds, collectively called ratites, and the tinamous. Many ratites are large and able to run fast and deliver powerful kicks when defending themselves from predators, but kiwis are small, shy forest birds that are generally nocturnal. Tinamous are New World birds that occupy a wide range of habitats. Tinamous can fly weakly. Paleognath birds share a primitive palate shape, from which they receive their scientific name, which means "old jaw."

● **Ratites** Ratites are a group of very large flightless birds with a long neck; thick, coarse feathering; powerful legs; and strong, thick toes. The word *ratite* relates to the flattened, raftlike shape of the sternum, or breastbone, which does not have the keel typical of birds. Most scientists agree that ratites evolved from flying ancestors, but it is unclear how closely the different groups of ratites are related to each other.

● **Kiwis** Living only in New Zealand, these bizarre-looking nocturnal birds have a cone-shaped body and a small head, and are covered with loose, rough-looking feathers. Kiwis have no tail and only tiny wings. Their long bill is used to probe for worms and insects on the forest floor, and their strong legs and feet with four long toes are used in digging for prey. DNA studies have revealed that the ancestors of these strange birds were very likely related to the emu and cassowaries. Kiwis are usually—but not always—considered to be ratites.

● **Emu** The emu is the largest bird in Australia and occurs nowhere else. It has loose, shaggy plumage, but the head and neck are bare. It has three toes and strong muscular legs. The emu can run at speeds of up to 30 miles an hour (50 km/h) over short distances. Its trachea (windpipe) has an unusual construction that enables the emu to make a very loud, booming call.

▶ *Only living species of palaeognaths are shown on this family tree.*

Animals
KINGDOM Animalia

Vertebrates
SUBPHYLUM Vertebrata

Birds
CLASS Aves

Ratities and tinamous
SUBCLASS Paleognathae

Tinamous
ORDER Tinamiformes

Kiwis
ORDER Dinornithiformes

Emus and cassowaries
ORDER Casuariiformes

Rheas
ORDER Rheiform●

Ostrich
ORDER Struthioformes

COMPARATIVE ANATOMY

Family ties

DNA analysis has revealed that kiwis are more closely related to the ostrich, emu, and cassowaries than to an extinct group of birds called the moas with which they once shared their isolated island home. This discovery has led scientists to conclude that the ancestors of modern ratite birds probably colonized New Zealand on two separate occasions.

● **Cassowaries** Cassowaries are the only living ratites that prefer dense forests to open country. The distinctive bony crest that tops the skull is covered with horn and may grow more than 6 inches (16 cm) tall in some males. Cassowaries have two fleshy crimson wattles, blue skin on their neck, and dark bristly-looking feathers covering the body. Although shy and retiring, cassowaries can be aggressive if provoked: their powerful legs can kick with sufficient force to kill a human, and an elongated toenail on each foot is capable of slicing open an attacker. The three species live in Australia, New Guinea, and a few adjacent islands.

● **Rheas** The New World counterparts of the ostrich and emu, rheas are smaller, with shaggy gray feathers covering most of a white or brown body and neck. They have three toes and live on the plains of southern South America from Brazil to Argentina.

● **Tinamous** There are nearly 50 species of tinamous, all living in South or Central America. They are ground-dwelling birds, and most live in forest habitats. More often heard than seen, tinamous can fly weakly. Tinamous are not ratites, and some scientists believe they are not paleognaths.

▲ *Sexually mature males vie with each other for the right to mate with females in the breeding season. The bold white plumes of feathers on the wings and tail identify these as males.*

● **Ostrich** The ostrich has proportionally tiny wings but enormous, powerful legs with two forward-pointing toes. The larger clawed toe is actually the third digit, and the smaller clawless toe is the fourth digit. Like most of its ratite relatives, the ostrich lives in open, often arid country, where its height and good eyesight give it an excellent view of the surroundings. It is the largest of the ratites and in nature is now confined to Africa. Former populations in the deserts of Syria and the Arabian Peninsula were hunted to extinction in the 1960s.

FEATURED SYSTEMS

EXTERNAL ANATOMY The ostrich is a huge flightless bird with large eyes, a very long neck, and powerful long legs that are covered in tough skin. *See pages 836–837.*

SKELETAL SYSTEM The ostrich has large bones, necessary for supporting its huge frame, but as in all birds the skeleton is comparatively lightweight. *See page 838.*

MUSCULAR SYSTEM Powerful muscles are necessary to support the ostrich's huge size and to provide strength as it sprints away from danger. *See page 839.*

NERVOUS SYSTEM The central part of the brain is smaller than it is in most birds, since the ostrich has no need for the sophisticated navigation systems associated with flight and migration. *See page 840.*

CIRCULATORY AND RESPIRATORY SYSTEMS The ostrich has nostrils located on the bill and a system of air sacs connected to the lungs that help to keep it cool in the heat of the African plains. *See page 841.*

DIGESTIVE AND EXCRETORY SYSTEMS The ostrich has a special stomach called a proventriculus that compensates for the absence of a gallbladder or crop such as that found in many other birds. *See page 842.*

REPRODUCTIVE SYSTEM Females lay up to 12 eggs, which are the largest of any bird, sometimes weighing 3 pounds (1,400 g). The young grow quickly after hatching. Unlike most birds, male ostriches have a relatively large penis. *See page 843.*

External anatomy

COMPARE the two huge toes on each foot of an ostrich with those of a *WOODPECKER*. Ostrich toes are adapted for supporting the weight of a heavy bird capable of running at speed. The woodpecker has four toes on each foot, with two facing forward and two backward. This arrangement allows the bird to climb vertically up tree trunks.

Ostriches are unmistakable birds, with a very long mobile neck, small head, large eyes, bulky feathered body, and incredibly long and powerful legs. Each foot has just two toes, which are covered with extremely tough skin. Ostriches are almost tailless. This species is a true record-breaker: as well as being the largest and heaviest living bird, it is also the fastest runner, capable of sprinting at up to 45 miles per hour (72 km/h). One bird was timed running at this speed for 20 minutes nonstop, during which time it traveled more than 15 miles. The ostrich's massive legs help it easily outrun most predators, and they can also

▶ **Male ostrich**
The dominant features of an ostrich are its long neck, two long and robust legs, and bushy tail.

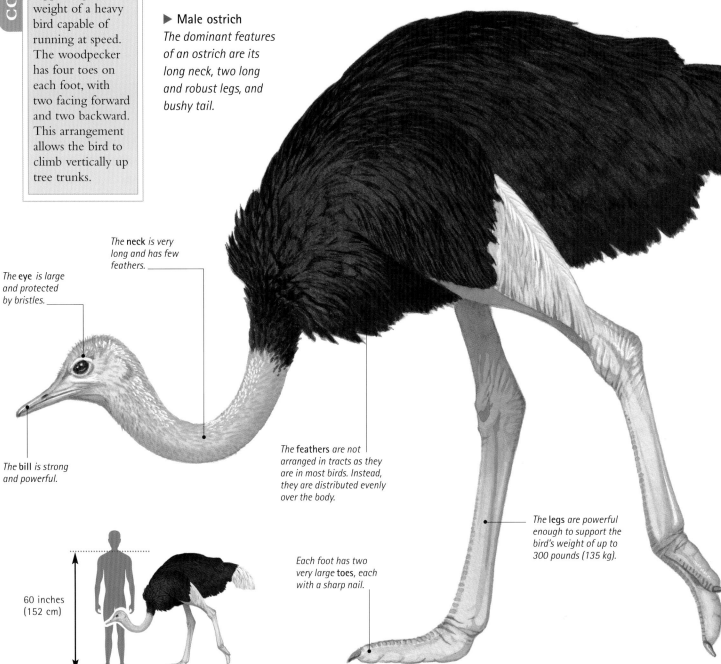

The **neck** *is very long and has few feathers.*

The **eye** *is large and protected by bristles.*

The **bill** *is strong and powerful.*

The **feathers** *are not arranged in tracts as they are in most birds. Instead, they are distributed evenly over the body.*

The **legs** *are powerful enough to support the bird's weight of up to 300 pounds (135 kg).*

Each foot has two very large **toes**, *each with a sharp nail.*

60 inches (152 cm)

become lethal weapons when the bird is ambushed: a kick from an adult ostrich can kill an enemy the size of a lion or a human. The ostrich also has the biggest eyeballs of any bird; they are 2 inches (5 cm) across and provide excellent vision.

Feathers and skin

Male ostriches are black with elaborate white plumes on the tail and at the wingtips. The skin on their head, neck, and legs is bright pinkish-red or blue. Females are much duller, with gray-brown feathers and skin, and the young are camouflaged with light brown stripes that disguise them from predators while they are small and vulnerable.

Ostriches are distinguished from other ratites by having just two toes, instead of three. As a result their feet look almost hooflike, a useful adaptation for running at great speed. The

▶ *The ostrich has the largest eyes of any bird. The pinkish areas around the eyes and on the throat are bare skin.*

The male ostrich displays the white feathers of its **tail** in courtship.

▼ FOOT
An ostrich's foot has only two toes, which are in fact the third and fourth digits. The other digits have been lost over the course of evolution.

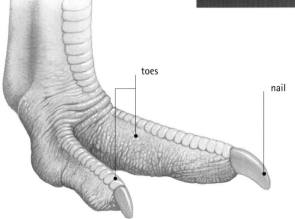

toes

nail

ostrich's unique toe formation is somewhat similar to the reduction in toes seen in running mammals of similar open habitats, such as gazelles and zebras.

Some ornithologists (scientists who study birds) believe that there may be two species of ostriches: the pink-skinned common ostrich, which is widely distributed from the Sahara to South Africa; and the Somali ostrich, which has blue skin and lives only in parts of southern Ethiopia, Somalia, and Kenya.

Skeletal system

CONNECTIONS

COMPARE the long, flexible neck of the ostrich with that of an *ALBATROSS*. The ostrich's neck allows it to take most of its food from the ground, whereas the albatross feeds mostly from the ocean surface.

All birds have a skeleton made of hollow, thin-walled bones. The ostrich is no exception, although its skeleton is more dense than that of flying birds. The spaces within the bones are filled with extensions of the body's air sacs, which supplement the lungs and contribute to a highly efficient respiratory system. With this structure, ostrich bones can grow very large and offer a bigger area for anchoring muscles and feathers without becoming enormously heavy.

The ostrich's skeleton has certain features not found in birds that can fly. The most significant feature is the lack of a keel-shaped sternum, so there is no point at which flight muscles can attach themselves to the breastbone. Instead, the ostrich has a very large, flattened breastplate that covers the entire chest area. This acts as a shield that protects internal organs such as the heart and liver.

The vertebral column is strong but flexible, especially along the neck; strength is necessary for supporting the ostrich as it feeds or runs. The first vertebra is called the atlas and, along with the second joint (the axis), supports the

▼ *An ostrich's sternum lacks the keeled shape typical of birds. The bones of the legs are the most robust of any bird species.*

The ear

Unlike mammals, birds have only a single bone—the columella—to transmit sound from the eardrum to the cochlea. Flying birds have specially adapted feathers surrounding the ear opening that minimize the effects of air turbulence. Ostriches lack these feathers.

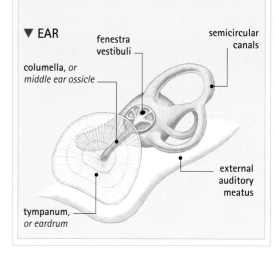

▼ EAR

- fenestra vestibuli
- semicircular canals
- columella, *or middle ear ossicle*
- external auditory meatus
- tympanum, *or eardrum*

skull while allowing the ostrich very flexible movement of its head and neck. Large eye sockets accommodate the huge eyes and reduce the weight of the skull.

Legs and wings

The structure of the leg and wing bones, including the splayed, horizontally positioned femur, is similar to that of other birds. However, the leg bones are unusually heavy for a bird; this heaviness is due to their sheer size and denser bone structure. Because ostriches live on the ground and have no need to perch on the branches of trees, they lack the opposable first toe bones of many flying birds. Instead, their feet have large reinforced bones suitable for running. The wing bones are relatively small, but their presence—along with the rigid backbone, a vestigial tailbone (the pygostyle), and certain associated muscles—is evidence that ostriches evolved from flying birds.

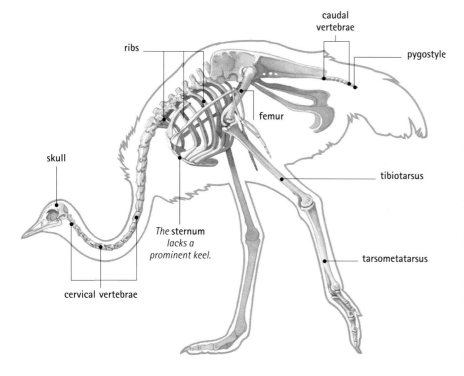

- caudal vertebrae
- ribs
- pygostyle
- femur
- skull
- tibiotarsus
- *The* sternum *lacks a prominent keel.*
- tarsometatarsus
- cervical vertebrae

Muscular system

The ostrich is a powerful runner, but it cannot fly. It has exceptionally large, strong leg muscles and small, weak flight muscles. Like all other ratites, the ostrich has no keel on its sternum, or breastbone. Without a keeled sternum, there is no major area for the attachment of the muscles that are essential for powering flight. An ostrich still has "flight" muscles, such as the pectoralis and the supracoracoideus muscles, and these muscles are evidence that in the distant past the ancestors of ostriches were able to fly.

Leg muscles

Each leg has a total of 36 muscles, located across five main areas: the pelvic, femoral, tibiotarsal, tarsometatarsal, and digital areas. Because an ostrich has only two toes on each foot, it lacks the muscles associated with other toes possessed by some birds.

The leg and thigh muscles of an ostrich have to be strong enough to deal with prolonged contractions as it runs. However, it is the foot, or tarsometatarsus—not the thigh—that takes most of the energy burden and absorbs the shock as the ostrich runs. A large amount of the energy needed during an extended sprint is stored elastically in the tendons, and the lower leg muscles contain great quantities of the pigment myoglobin, which stores the oxygen needed to prevent muscular fatigue. Myoglobin gives the muscles in that area a darker coloration.

The largest leg muscle, the gastrocnemius, is the major source of running power and has a far greater capacity for bursts of high energy than the closely associated digital flexor muscles. These more flexible muscles are probably more important for maintaining balance and for the storage and release of tendon energy.

COMPARE the supracoracoideus "flight" muscles of the ostrich with those of a **PENGUIN**.

COMPARE the structure of the leg muscles in an **OSTRICH** with those of a **SLOTH**.

CONNECTIONS

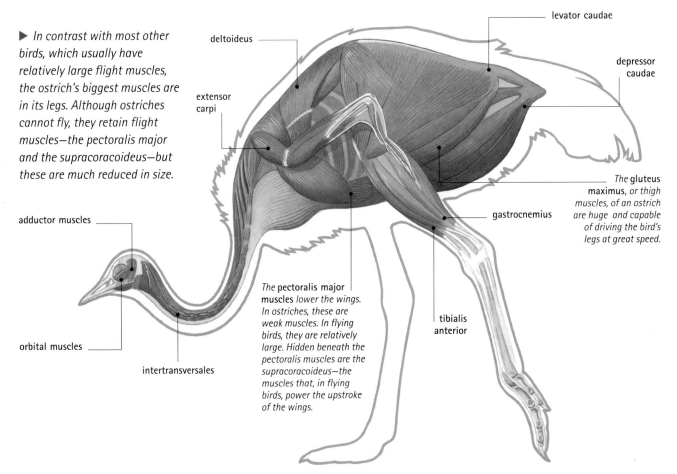

▶ In contrast with most other birds, which usually have relatively large flight muscles, the ostrich's biggest muscles are in its legs. Although ostriches cannot fly, they retain flight muscles—the pectoralis major and the supracoracoideus—but these are much reduced in size.

levator caudae

deltoideus

depressor caudae

extensor carpi

The gluteus maximus, or thigh muscles, of an ostrich are huge and capable of driving the bird's legs at great speed.

gastrocnemius

adductor muscles

The pectoralis major muscles lower the wings. In ostriches, these are weak muscles. In flying birds, they are relatively large. Hidden beneath the pectoralis muscles are the supracoracoideus—the muscles that, in flying birds, power the upstroke of the wings.

tibialis anterior

orbital muscles

intertransversales

839

Nervous system

CONNECTIONS

COMPARE the cerebellum of the ostrich with that of a flying, migratory bird such as the *ALBATROSS*.

COMPARE the cerebral cortex of the ostrich with that of the *CHIMPANZEE*.

COMPARE the spinal cord of the ostrich with that of the *GULPER EEL*.

The ostrich's central nervous system (CNS) comprises the brain and spinal cord. The CNS works in the same way as that of other birds and mammals; its main functions are to process and centralize sensory impulses from the bird's environment, coordinate voluntary and involuntary movements and functions, and store information.

The spinal cord is part of the CNS. It is made up of nerve cells and is sheathed and protected by the vertebral column. The cord carries 31 pairs of spinal nerves that connect to the peripheral nervous system.

The ostrich's brain, like that of all birds, is dominated by the central part of the cerebral hemisphere. In mammals the dominant part is the top layer, or cerebral cortex, which provides an increased ability to learn and store new information. The cerebellum controls the coordination of movement. It is smaller in ostriches than in flying birds. A third area—the cerebrum—controls reproductive physiology and behavior such as mating and caring for eggs and chicks.

▼ *A complex system of nerve fibers and tissues enables the ostrich to process sensory impulses from its environment, coordinate movement, and store information.*

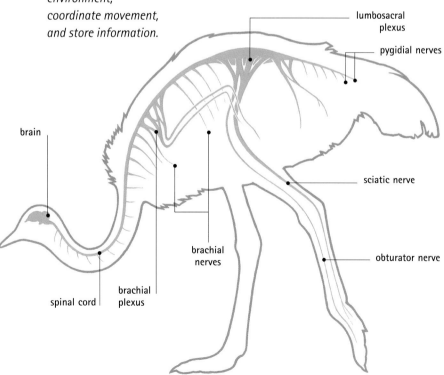

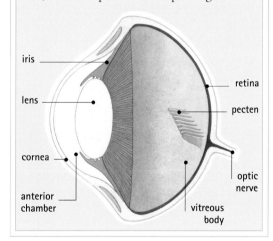

IN FOCUS

Eyesight

At 2 inches (50 mm) in diameter, the ostrich eye is the largest of any land bird or mammal. The position of the eyes on the side of the head gives the bird a wide field of vision. The general structure of the eye is the same as in most birds and mammals, and the lens is very flexible, enabling the ostrich to change its focus rapidly from near to far objects when scanning the horizon for potential threats. The ostrich retina (the light-sensitive area at the back of the eye) is thicker than that found in mammals, and has far more light-sensitive cells (rods and cones). Because the ostrich is active by day, its retina has a high density of cells called cones, which process colors and bright light. Nocturnal birds such as owls have retinas packed with rods, which help them see in poor light.

Ostrich senses

The optic lobes of the ostrich's brain are large, and the olfactory lobes are small, reflecting the relative importance of vision over smell. Flightless birds tend to have a better sense of smell than their flying counterparts, but the ostrich is still no match for its relatives, the kiwis. Kiwis are the only birds with nostrils at the end of the bill. Kiwis actively sniff out their prey in forests at night.

Circulatory and respiratory systems

The circulatory system is responsible for the transport of food, gases, hormones, and waste products. The ostrich has a large, powerful heart with four chambers. The right side of the heart pumps oxygen-depleted blood from the body cells to the lungs. The left side then receives oxygenated blood from the lungs and pumps it to the body cells. The ostrich has a heart rate similar to that of a human: about 70 beats per minute. The heart rate is slow compared with that of smaller, more active birds; a chickadee's heart, for example, may beat up to 1,000 times per minute.

The ostrich has one large jugular vein that runs down the right side of the neck, and this is strengthened to cope with the greater pressure needed to pump blood such a long way against gravity. Throughout the circulatory system, veins and arteries lie next to each other, allowing the warm blood coming from the heart to warm the cooler blood draining back from the extremities. However, unlike birds that live in colder climates, ostriches do not need to conserve a great deal of heat at the body core, so they lose more heat from the skin. Blood returning through the veins from the appendages is far cooler.

Respiration

The respiratory system of the ostrich, as with all birds, consists of the lungs and a system of air sacs that begin in the thorax and reach into some of the bones. The windpipe, or trachea, is large and opens in the lower part of the mouth. The esophagus opens above the trachea

IN FOCUS

Keeping cool

The ostrich has a specially adapted system of air sacs to help to keep its temperature down during the intense heat of the day. Body heat is reduced by panting, causing the air sacs connected to the lungs to inflate and deflate. The lungs are far more rigid than those of mammals; however, the air sacs provide plenty of spare capacity. The normal respiratory rate of an ostrich is low—between 7 and 12 breaths per minute—but this increases when the bird is running.

COMPARE the ostrich's heart with that of a *HUMMINGBIRD*.

COMPARE the lungs of the ostrich with the gills of a *SAILFISH*. Both organs are means to draw oxygen into the body.

CONNECTIONS

and runs down the right side of the neck. Ostriches have a slower respiratory system than many birds, owing to their great size, slower metabolic rate, and flightlessness. They have no need for the supercharged system required by long-distance migrants or other birds that fly at altitudes where the air is thin.

▶ **CIRCULATORY SYSTEM**
A network of arteries and veins connects the heart and lungs with all parts of the body. As is the case with all other birds, the ostrich has two lungs and a system of air sacs for respiration.

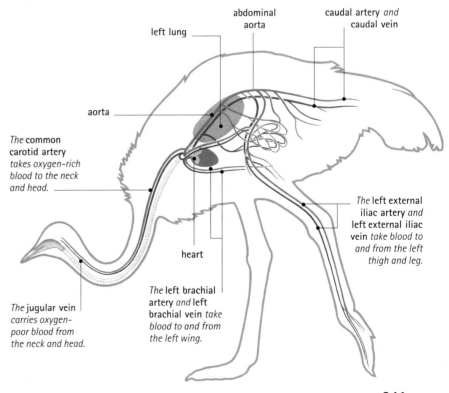

abdominal aorta

caudal artery *and* caudal vein

left lung

aorta

The **common carotid artery** *takes oxygen-rich blood to the neck and head.*

The **left external iliac artery** *and* **left external iliac vein** *take blood to and from the left thigh and leg.*

heart

The **left brachial artery** *and* **left brachial vein** *take blood to and from the left wing.*

The **jugular vein** *carries oxygen-poor blood from the neck and head.*

Digestive and excretory systems

COMPARE the digestive tract of the ostrich with that of a ruminant mammal such as the **RED DEER**. Despite a similar diet, the ostrich has a single-chamber stomach, but the red deer's stomach has four chambers.

COMPARE the cloaca of the ostrich with the excretory apparatus of a bivalve, such as a **GIANT CLAM**.

Ostriches feed mostly on large quantities of plant matter such as seeds, grasses, flowers, and succulent plants. Occasionally, they will also catch small reptiles and insects such as locusts to supplement their diet, and they sometimes—though rarely—eat carrion left by predators.

The digestive tract

The ostrich lacks the crop found in many birds. Food passes directly from the esophagus into a stomach called the proventriculus. There, enzymes such as pepsin and other digestive juices go to work on the food, which then passes into a second, muscular stomach called the ventriculus, or gizzard. Usually, the ventriculus contains stones and grit that the ostrich swallows to help grind tough plant material into pulp. Once the food has been ground, pulp passes into the small intestine a little at a time via the pyloric sphincter. The first section of the small intestine is the duodenum. There, bile from the liver and pancreatic juices are introduced to help break down fats and promote the absorption of nutrients into the bloodstream.

CLOSE-UP

A high-fiber diet

The ostrich eats large quantities of vegetation rich in fiber and needs an efficient digestive system to process this effectively. The fiber is digested only when it passes into the large intestine and cecum, where millions of microbes help break it down. Without the gut microbes, the bird would be unable to digest its food properly. Such large quantities of fiber provide the ostrich with a rich source of volatile fatty acids, essential for energy and building muscle tissue.

The ostrich's small intestine is unusually long, so food takes up to 36 hours to pass through it. The system allows plenty of time for the ostrich's body to extract the maximum nourishment. Waste products from the digestive and urinary tracts are discharged from the cloaca, as in most birds.

▶ The ostrich's digestive tract is long, enabling it to extract the maximum amount of nourishment from its food. Nevertheless, the ostrich must eat large quantities of food to sustain itself.

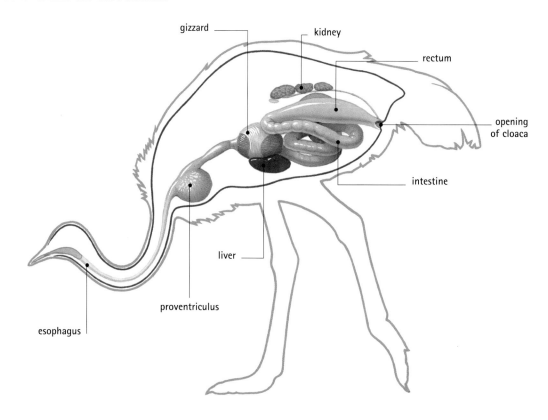

Reproductive system

The reproductive tract of the female ostrich consists of a single tract and ovary with a small clitoris located on the ventral cloaca. The male has two testes, which produce sperm, and a penis for delivering it to the female during mating. Few species of birds have a fully formed penis. The ostrich penis does not contain the urethra.

Before ostriches mate, they participate in elaborate courtship rituals, in which territorial males compete for small flocks of females. After dividing into mating groups, ostriches in some regions create communal nests where females may gather to lay up to 60 eggs.

Egg production

Ostriches lay the biggest eggs of any living birds, with average measurements of 4.5 by 7 inches (11 by 18 cm) and weighing 3 pounds (1,400 g). Single females lay up to 12 huge whitish eggs in a specially created nest scrape.

The ostrich egg begins as a single cell. It is produced initially within the female's ovary and is transferred to the oviduct, where it is fertilized by the male's sperm. From there it passes along the oviduct, becoming coated with albumen (the viscous protein that forms egg white). The egg is then wrapped in the chorion membrane, which forms the papery lining of the inner eggshell. Farther along, the lining of the duct produces secretions that form the hard outer shell. The egg is fully formed and ready to be laid about 24 hours after it leaves the ovary. By then the embryo is

Inside the egg

As the ostrich embryo develops in the egg, it is nourished via membranes that grow out of its digestive tract and contact the surrounding yolk. These membranes produce digestive enzymes which dissolve proteins in the yolk, and the yolk sac develops a network of tiny blood vessels that transport the food directly to the embryo. Metabolic waste produced by the embryo is stored in a membrane called the allantois.

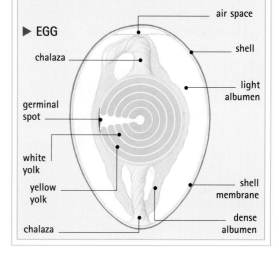

▶ EGG

air space
shell
light albumen
shell membrane
dense albumen
chalaza
germinal spot
white yolk
yellow yolk
chalaza

visible as a pale disk on the surface of the yolk. Further development of the embryo depends on the egg's being kept at the correct temperature during incubation.

The incubation period lasts six weeks, and although both parents help protect the eggs, fewer than 10 percent survive to hatching. The downy young are able to see, move around, and regulate their body temperature within minutes of hatching (they are precocious). The young reach full adult height within a year.

STEVEN SWABY

FURTHER READING AND RESEARCH
Proctor, Noble S., and Patrick Lynch. 1993. *Manual of Ornithology.* Yale University Press: New Haven, CT.

▼ MALE CLOACAL REGION
The ostrich is one of the few birds with a fully formed penis. During mating, the penis emerges from the entrance of the cloaca. The cloaca has three chambers: the coprodeum, urodeum, and proctodeum.

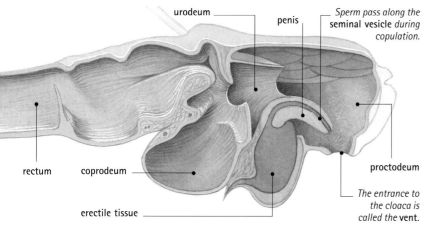

urodeum
penis
Sperm pass along the seminal vesicle *during copulation.*
rectum
coprodeum
erectile tissue
proctodeum
The entrance to the cloaca is called the **vent.**

Otter

ORDER: Carnivora FAMILY: Mustelidae
SUBFAMILY: Lutrinae

The 13 species of otters live in and around rivers, estuaries, and sea coves. Otters are adept swimmers and hunt for most or all of their food in the water. The giant otter and the eight species of river otters feed mainly on fish. In the other four species, shellfish make up the bulk of the diet.

Anatomy and taxonomy

Scientists categorize all organisms into taxonomic groups based on their anatomical, biochemical, and genetic similarities and differences. The 13 living species of otters are mustelids, members of the weasel family, which also includes skunks, badgers, and honey badgers.

● **Animals** Otters, like other animals, are multicellular and obtain their energy from eating other organisms. Animals differ from other multicellular life-forms in their ability to move from one place to another (in most cases, using their muscles and skeleton). They generally react rapidly to touch, light, and other stimuli.

● **Chordates** At some time in its life cycle, a chordate has a stiff, dorsal (back) supporting rod called the notochord that runs all or most of the length of the body.

● **Vertebrates** In vertebrates, the notochord develops into a backbone (spine) made up of units called vertebrae. The vertebrate muscular system that moves the head, trunk, and limbs consists primarily of muscles arranged in mirror-image on either side of the backbone (bilateral symmetry).

● **Mammals** Mammals are warm-blooded vertebrates that have hair made of keratin. Females have mammary glands that produce milk to feed their young. In mammals, the lower jaw is a single bone (the dentary), which is hinged directly to the skull—a different arrangement from that found in other vertebrates. A mammal's inner ear contains three small bones (ear ossicles), two of which evolved from the jaw mechanism of reptilian ancestors. Mammalian red blood cells, when mature, lack a nucleus; all other vertebrates have red blood cells that contain a nucleus.

▼ This family tree shows the major groups to which otters belong. Biologists recognize 13 living species of otters, which are classified in three groups: lutra-types, lontra-types, and the giant otter.

Animals
KINGDOM Animalia

Vertebrates
SUBPHYLUM Vertebrata

Mammals
CLASS Mammalia

Placental mammals
SUBCLASS Eutheria

Marsupials
SUBCLASS Metatheria

Carnivores
ORDER Carnivora

Bears
FAMILY Ursidae

Cats
FAMILY Felidae

Mustelids
FAMILY Mustelidae

Dogs, Foxes, and Wolves
FAMILY Canidae

Seals
FAMILY Phocidae

Otters
SUBFAMILY Lutrinae

Lontra-type otters

Lutra-type otters

Giant otter
GENUS AND SPECIES
Pteronura brasiliensis

North American river otter
GENUS AND SPECIES
Lontra canadensis

Sea otter
GENUS AND SPECIES
Enhydra lutris

Eurasian river otter
GENUS AND SPECIES
Lutra lutra

Number of otter species

Between the 1960s and 1980s, different biologists identified between 9 and 19 species of otters in total. Part of the problem in identifying different types of otters lies in the great variability of characteristics within a species. Some distinguishing features—such as tooth size and the hairiness of the nose—show geographic variation in populations of the same species. Thus these features are not reliable for distinguishing species. In the 1990s, the use of biochemical data, especially DNA analysis, enabled otter specialists to decide upon 13 species of otters. However, many relationships between otter species remain unresolved; for example, little is known about the spotted-necked otter, the smooth otter, and the giant otter.

▲ Streamlined for fast, efficient swimming, the North American river otter preys on fish and is found in rivers and lakes throughout Canada and much of the United States.

● **Placental mammals** These mammals nourish their unborn young through a placenta, a temporary organ that forms in the mother's uterus (womb) during pregnancy.

● **Carnivores** The word *carnivore* is sometimes applied to animals that eat meat, but it applies more specifically to members of the mammalian order Carnivora. Members of this group include dogs, cats, raccoons, mustelids, civets, bears, and hyenas. Most members eat meat almost exclusively, but some, such as badgers and most bears, have a mixed diet; and a few, such as the giant panda, eat only plants.

● **Mustelids** The mustelids, or the weasel family, range in size from the least weasel, weighing a mere 2 to 3 ounces (56-84 g), to the wolverine, sea otter, and giant otter, all of which weigh more than 70 pounds (32 kg). Mustelids have an elongated or stocky body with short legs. All species have small ears, and five toes on both their fore- and hind feet (most other carnivores only have four toes on the hind foot). The claws are typically long, sharp, and nonretractable.

● **Otters** Although the mink—a weasel-like carnivore—often hunts in freshwater, the 13 species of otters are the only mustelids with a body suited to life in water. Otters are comfortable in water or on land, except for the sea otter, which is rarely ashore. Until the 1990s, the classification of otters was largely based on structural features, such as the shape of the teeth and the male genitals. Recently, DNA analysis has led to a revision of classification. Otters are now placed in three groups: the *Lontra*-type otters of the New World; the *Lutra*-type species, mostly of the Old World; and the giant otter (*Pteronura*) of South America.

FEATURED SYSTEMS

EXTERNAL ANATOMY Otters have an overall body shape typical of members of the weasel family. They are streamlined, with a long tapering tail and partially webbed feet for locomotion in water. *See pages 846–849.*

SKELETAL SYSTEM The otter's skeleton is a compromise for movement both on land and in water. *See pages 850–852.*

MUSCULAR SYSTEM Otters have strong leg and tail muscles that help power them through water. *See page 853.*

NERVOUS SYSTEM The brain of the otter is similar to that of other mustelids. Sight and touch become the major senses when the otter is underwater. *See pages 854–855.*

CIRCULATORY AND RESPIRATORY SYSTEMS Otters do not dive very deep or for long. Sea otters divert blood to the surfaces of their feet to lose heat to their surroundings, thus helping regulate body temperature. *See pages 856–857.*

DIGESTIVE AND EXCRETORY SYSTEMS Otters have a simple gut, like other carnivores, but it is unusually long to cope with the high food intake needed to maintain body temperature. *See pages 858–859.*

REPRODUCTIVE SYSTEM The sea otter and the North American river otter can delay implantation of a fertilized egg and so prolong the gestation period, allowing young to be born in more favorable conditions. *See pages 860–861.*

External anatomy

COMPARE the vibrissae (whiskers) of otters with those of *SEALS*. In both animals, the vibrissae are sensitive to movement and vibration beneath the water surface.

CONNECTIONS

The streamlined body of all otter species enables them to swim quickly and efficiently underwater. When swimming slowly on or below the water surface, an otter "dog-paddles" using all four limbs, almost as though it were walking through water. When it is swimming fast, the ears, front legs, and body hairs fold back against the body, giving the otter a torpedo-like shape. The tail is typically about one-third of total body length and is slightly flattened from top to bottom. The tail serves as both a rudder for steering and a paddle, which sculls up and down to propel the otter forward underwater at high speed. When the otter is swimming fast, the webbed hind limbs are extended and move in unison with the tail. On land, the tail helps the otter balance when walking, and when it stands up on its

▼ Sea otter
Many of the otter's external features are suited to life swimming and hunting underwater.

Vibrissae (whiskers) detect prey in conditions of poor visibility.

Otters have **dense fur** with two layers: longer guard hairs and shorter underfur for insulation.

small ear

length up to 52 inches (1.35 m)

gripping pad

retractable claw

A webbed **hind foot** *helps the otter swim fast and powerfully.*

Otter's ancestry

The earliest recognizable otterlike creatures in the fossil record date back to about 30 million years ago. They had sharp carnassial (cheek) teeth for slicing, which suggest that their diet was predominantly fish. Some later forms have carnassial teeth that evolved to become more flattened for crushing, similar to modern forms such as the sea otter. This development suggests that some otters turned to eating mainly aquatic invertebrates only later on. The features of modern otters suited to an aquatic way of life are modifications of features that are also found in ancestral carnivores, rather than features unique to otters. For example, thick fur and webbed feet are more pronounced in otters than in other carnivores, but even dogs have thick fur, have partially webbed feet, and can swim relatively well.

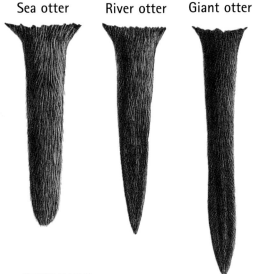

Sea otter　　River otter　　Giant otter

▲ OTTER TAILS

Tails vary in size and shape from species to species. The sea otter's tail is relatively short—about one-quarter of its body length. In river and giant otters, it is relatively longer—about one-third of body length.

hind legs to get a wider view it leans back on its tail for support.

The sea otter has particularly large webbed hind feet, which produce a paddle area some five times greater than that of a river otter of an equivalent size. The greater reliance on hind limbs for propulsion may help explain the sea otter's relatively short tail—only about one-quarter of the total body length. Sea otters are also unusual in that they commonly swim on their back. In this position, they propel themselves by sculling the tail from side to side.

Alternatively, they use their hind limbs for propulsion, either sculling in unison like a human rower or sculling first on one side and then the other, like a paddling canoeist.

Otter fur varies from brown or brownish gray on the back to slightly paler on the throat and belly. Otter cubs start life with gray fur. The brown fur of mature otters may become grizzled with age.

Otter fur has the highest density of hairs of any mammal—commonly at least 450,000 hairs per square inch (70,000 per sq cm). The glossy, dense fur—highly valued by humans for its attractive appearance and insulating properties—is one reason why, until the 20th century, otters were heavily hunted by humans. The combination of tightly packed underfur, together with longer guard hairs, traps a layer of air above the surface of the skin, creating the otter equivalent of a wet suit. The trapped air acts as insulation, reducing heat loss from the body while providing buoyancy. Otters groom their fur by licking and stroking it with their forepaws. This ensures that oil secreted from sebaceous glands in the skin is spread over the hairs to keep them waterproof. The grooming also introduces air into the fur and prevents it from becoming matted.

The thickness of an otter's fur pelt depends on species and local conditions. Generally, it

*Short **tail** used as a rudder and paddle during swimming.*

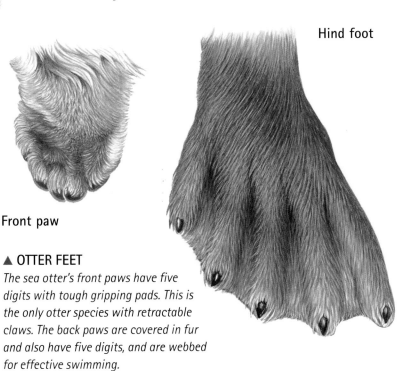

Front paw

Hind foot

▲ OTTER FEET

The sea otter's front paws have five digits with tough gripping pads. This is the only otter species with retractable claws. The back paws are covered in fur and also have five digits, and are webbed for effective swimming.

American river otter

Eurasian otter

Sea cat

Giant otter

◀ OTTER NOSE PADS

The hairiness of an otter's nose pad varies between species, increasing as the otter's habitat moves nearer the equator. Nose pads range from hairless in the American river otter to very hairy in the giant otter.

reflects the amount of time the otter spends in water. Sea otters spend their lives in cool, temperate, or subpolar waters and have the thickest fur. Unlike other aquatic mammals such as seals, whales, and sea cows, otters do not lay down a layer of fatty blubber beneath the skin as insulation. For this purpose, the air trapped within their fur suffices. A 1-inch (2.5-cm) thickness of air provides as much insulation as the 4-inch layer of blubber found in seals and whales. Except for the sea otter, which has been recorded diving to depths of about 330 feet (100 m), otters are not deep divers. However, even at a depth of 33 feet (10 m), the pressure of water compresses the insulating layer of air to half of its thickness, thus halving its insulating effectiveness. Otters dive for only short periods compared with other aquatic mammals, and they lose relatively large amounts of body heat underwater. To compensate, they have a high metabolic rate (they "burn" large amounts of food to release energy to keep themselves warm), and so they must consume 15 to 25 percent of their body weight in food each day.

The head of the otter is broad with a fairly blunt muzzle and ends in a black nose pad that ranges from hairless in the American river otter to completely covered with hair in the giant otter and the hairy-nosed otter. The eyes and

COMPARATIVE ANATOMY

Streamlining

Fully aquatic mammals, such as porpoises and manatees, are streamlined to reduce drag as they swim underwater. Freshwater otters, by contrast, spend part of their time on land and in burrows, and sea otters spend much of their time floating on the sea surface. The shape of the otter, therefore, is a compromise for moving effectively on land or at the water surface in addition to under the water, and is less streamlined than that of fully aquatic animals. Porpoises and manatees have lost their hind limbs, whereas hind limbs have been retained in otters and seals, allowing them to move about on land.

nostrils are located toward the top of the head, enabling the otter to see above the water and breathe air with most of its head submerged. The otter's earflaps are small—typically less than 1 inch (2.5 cm) long—and a valve closes the outer ear when the otter dives. Otters have extremely good hearing in air, but when they dive they must use senses other than hearing to locate food and avoid predators. Like other mustelids, otters also have an acute sense of smell in air, but valvelike flaps close their nostrils when they submerge. Sight and touch come to the fore when otters dive beneath the water.

Stiff, long, touch-sensitive hairs called vibrissae form whiskers on either side of the snout. Similar hairs also grow on the otter's

glossy fur

broad muzzle

muscular tail

◀ Eurasian otter

Native to Europe, Asia, and parts of North Africa, the Eurasian otter grows up to 44 inches (110 cm) long and has brownish gray to brown fur on its back, with paler fur on its underside.

Otter sizes

The smallest otter species is the Asian short-clawed otter, which usually measures less than 36 inches (90 cm) from head to tail, and weighs up to 12 pounds (5.5 kg). The Eurasian otter and the North American river otter are of intermediate size. The giant otter and sea otter are the largest, reaching lengths of 6 feet (1.8 m) and weighing up to about 70 pounds (32 kg).

▼ *The Asian short-clawed otter is the smallest species of otter. It also has a slightly rounder head and body shape compared with other otters.*

Otter hair

Otter skin is riddled with thousands of hair canals. Emerging from each hair canal is a single long, coarse guard hair and several dozen shorter, soft underfur hairs that grow from hair follicles. The guard hairs help protect the soft underfur, and as they dry and straighten, open up the underfur, allowing air to enter and provide insulation. Sebaceous glands in the skin secrete an oily fluid that waterproofs the hair.

clawless otter—the forepaws have fingers that detect prey items by touch. When an otter is moving on land, most of its weight is carried by the hind limbs, and these provide the main thrust when the otter swims slowly. They also supplement the tail's propulsive force during rapid swimming. On land, the otter walks, runs, or jumps in a rather awkward, weasel-like fashion, with its back hunched. The sea otter, the species most suited for life in water, walks quite clumsily on land and does not run.

elbows. Underwater, the otter uses its vibrissae to help navigate by touch and to detect the swirls and vibrations produced by prey and predators moving in the water.

In good light, otters can see moderately well both above and below the water. Air and water have different light-bending properties, and the lens in the otter's eye changes shape in order to compensate for this difference. The lenses bend light rays much more when the otter is underwater than when it is looking through air.

Otter forelimbs tend to be shorter and less muscular, with the paws less webbed, than their hind limbs. Otters use their forepaws for manipulating food items and as paddles in slow swimming. In the two clawless species of otters—the Cape clawless otter and the Congo

▶ MANIPULATING FOOD

Sea otters have a pouch of skin under each front leg that extends over the chest. The pouch has room for food, usually mollusks and crustaceans, and a stone, which the otter uses to break open food items.

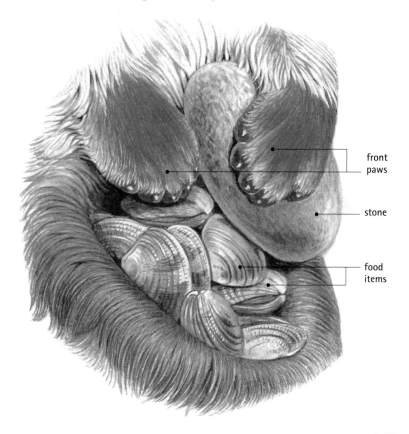

front paws

stone

food items

Skeletal system

COMPARE the fore- and hind limbs of an otter with those of a *SEAL*. Both are aquatic mammals that spend part of their time on land. Their limbs permit locomotion in water and on land.

As in other species of vertebrates, the otter's skeleton shapes and supports the body, protects vital internal organs such as the brain, heart, liver, and lungs, and serves as a point of attachment for skeletal muscles that move parts of the body.

The skeleton of an otter is based on a plan similar to that of other carnivores, especially members of the weasel family, but it has modifications for an aquatic way of life. Otters are highly flexible. This flexibility helps them twist and turn easily when they hunt prey, and allows them to reach all parts of their pelt for grooming, thus ensuring that the fur remains waterproof and that the underfur remains air-filled for insulation.

Skull and jaw

Otters, like most other carnivores, slice, chew, and crush the protein-rich food in their fleshy diet. They tend to swallow chunks of food, rather than grind it into a soup. The hinge joint between lower jaw and skull is quite tight, allowing up-and-down movement for grasping prey and slicing flesh but not side-to-side movement for grinding.

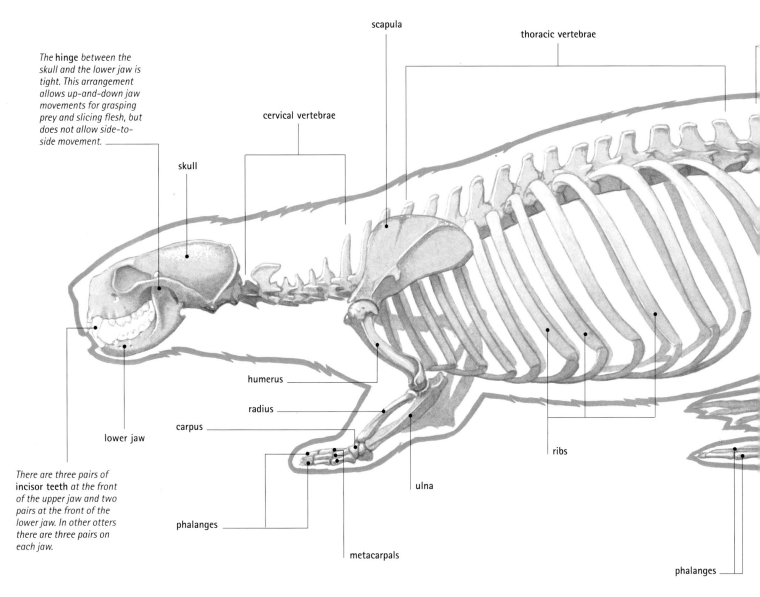

The **hinge** between the skull and the lower jaw is tight. This arrangement allows up-and-down jaw movements for grasping prey and slicing flesh, but does not allow side-to-side movement.

There are three pairs of **incisor teeth** at the front of the upper jaw and two pairs at the front of the lower jaw. In other otters there are three pairs on each jaw.

scapula

thoracic vertebrae

cervical vertebrae

skull

lower jaw

humerus

radius

carpus

phalanges

metacarpals

ulna

ribs

phalanges

► SKULL AND TEETH

The sea otter has a long, flat skull, streamlined like the rest of its skeleton. Its incisors are used for scooping invertebrate prey from their shell. The canines are long for piercing and ripping, and the carnassial teeth are flattened for crushing and grinding shells (other otters have sharp carnassial teeth for tearing flesh).

eye socket

zygomatic arch

flat skull

canines

incisor

carnassial teeth

mandible

lumbar vertebrae

sacral vertebrae

femur

tibia

fibula

calcaneus

There are 20 or 21 caudal vertebrae.

tarsals

metatarsals

◄ **Sea otter**

Major bones of the sea otter. Otters have a highly flexible skeleton, allowing them great agility in water. Otters' limbs are suitable for movement both on land and in water, but in sea otters the bones of the hind-limb toes are particularly long, making the hind paws effective paddles for swimming.

Like other carnivores, otters have three pairs of incisors at the front of each jaw—except sea otters that have only two pairs of lower incisors. Otters use their incisor teeth for grooming. The teeth serve as combs to remove tangles and trapped particles from their fur. In sea otters, the enlarged lower incisors also act as spadelike scoops to extract flesh from the shells of invertebrate prey.

In common with most other carnivores, otters have long, stabbing canines for grasping prey. The carnassial teeth of otters range between two extremes. In most river otters, they are sharp, with a shearing action for slicing fish flesh. In sea otters and African clawless otters, the carnassial teeth are flattened for crushing and grinding hard-shell invertebrates. Other otters have carnassial teeth shaped somewhere between these two extremes, a compromise between slicing and crushing.

Backbone

The sea otter's backbone typically has 50 to 51 vertebrae: seven cervical (neck), 14 thoracic (chest), three sacral (pelvic region), and 20 to 21 caudal (tail) vertebrae. The backbone of otters, in general, supports limbs that are used not only for swimming but also for walking or

lumbar vertebrae

sacral vertebrae

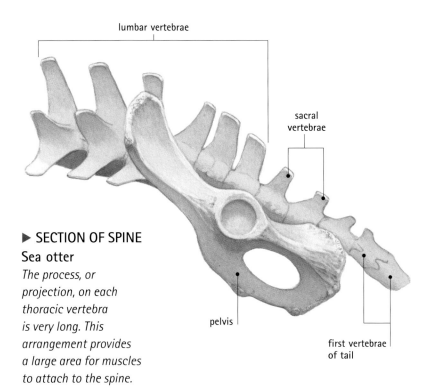

▶ SECTION OF SPINE
Sea otter
The process, or projection, on each thoracic vertebra is very long. This arrangement provides a large area for muscles to attach to the spine.

pelvis

first vertebrae of tail

running on land. Thus the backbone is similar to that of land carnivores, but with some modifications for an aquatic lifestyle. For example, all otters swim rapidly underwater by moving their tail, hind feet, and pelvic region up and down together.

Limbs and their supports

As in other mammals, the limbs of otters are connected to the spine through two limb girdles. The front or pectoral (shoulder) girdle contains two scapulae (shoulder blades) that connect the upper bone, the humerus, of each forelimb to the backbone through ligaments and muscles. As in some other carnivores, the pectoral girdle of otters lacks clavicles (collarbones)—a feature typical of mammals that move the forelimbs mostly in one plane, such as when the animal is running or paddling. The lower forelimbs of otters are extremely flexible, permitting various types of movement for walking, jumping, and paddling, and for grasping and handling food.

In otters, as in land mammals generally, the rear girdle (the pelvic girdle or pelvis) is anchored to the backbone through sacral vertebrae that are fused. The hind limbs of otters—carrying most of the body weight on land and forming major paddles in the water—are larger than the forelimbs. The hind

limbs of sea otters have taken their role as paddles one stage further. The bones of the hindfeet that make up the "toes" (the metatarsals and phalanges) are elongated, and the length of the toes is unusual in that the outer toe is longest and the inner toe is shortest—the reverse situation to that found in many mammals. In sea otters, the outer toe forms the leading edge of the paddle and ensures the animal scoops plenty of water with each stroke. Having the hind feet in this arrangement makes the sea otter "flat-footed" on land—its "ankle" bones, the tarsals, join the metatarsals and phalanges in forming a large pad in contact with the ground. This is called the plantigrade stance and makes the sea otter a slow walker on land. The Eurasian otter, by contrast, has its ankle bones raised off the ground; this is the digitigrade stance. This arrangement effectively increases the length of the legs and gives the animal much greater mobility on land.

COMPARATIVE ANATOMY

Forelimbs

Differences in feeding strategy help account for anatomical differences in the forelimbs of different otter species. Most river otters actively hunt fish and grasp the prey initially in the mouth. They have long claws on their forelimbs to help them handle the slippery, wriggling prey when eating it. The African clawless otters and the Asian short-clawed otter eat mainly shellfish, which they find and collect by feeling along riverbeds or coastal seabeds. They have sensitive front paws with fingerlike digits. The sea otter, too, has sensitive front paws, but, unlike other otters, it has retractable claws (they can be sheathed away like those of a cat). The sea otter's toes are covered with loose, hairy skin—almost like a mitten—and there are tough pads on the palm that help it grasp spiny or slippery invertebrates. The sea otter is one of the world's few tool-using animals. It breaks open the shells of shellfish—crabs, and mollusks such as mussels and abalone—by floating on its back and pounding the prey against a rock balanced on its chest.

Muscular system

Like other vertebrates, otters have three types of muscles. Skeletal, or striated, muscles are attached to bones and are used for moving parts of the body, locomotion, and maintaining posture. Smooth, or involuntary, muscles are present in many internal organs and structures, such as the intestines and blood vessels. The smooth muscle of the intestine helps move food along the digestive system. Cardiac muscle is a specialized type of muscle that is only in the walls of the heart and pumps blood around the circulatory system.

Muscles of the head

In general, the muscular system of otters is similar to that of other mustelids, such as badgers, honey badgers, and weasels. Like most carnivores, otters have powerful masseters (cheek muscles). These give them the strength to slice fish or crush the shells of invertebrate prey. Otters' lips, too, are particularly muscular, and they use these to manipulate items of food.

The eyes have strong ciliary and iris muscles, which control the curvature of the lens. Underwater, these muscles contract, making the lens bulge out. This mechanism allows the otter's eye to focus clearly underwater. In addition, muscles on the head close off the ears and nose when the otter dives.

Epaxial power

Otters have a muscular neck and streamlined body. The limbs are stout and heavily muscled, especially the hind legs, which, along with the slightly flattened, muscular tail, are used for rapid swimming.

A major difference between otters and their land-living relatives is the size of the epaxial muscles. These sets of muscles—which move the pelvic and tail regions in the vertical plane—are much larger in otters than in land-living mustelids. The powerful epaxial muscles allow the otter's tail to serves as both a rudder for steering and a paddle, propelling the otter rapidly through the water. Correspondingly, the neural spines and transverse processes on the lumbar, sacral, and caudal vertebrae to which the epaxial muscles are attached are also significantly larger than in their land-living relatives.

▼ SUPERFICIAL MUSCLES
Sea otter
The otter is a sleek, muscular animal with strong leg and tail muscles to power it through water.

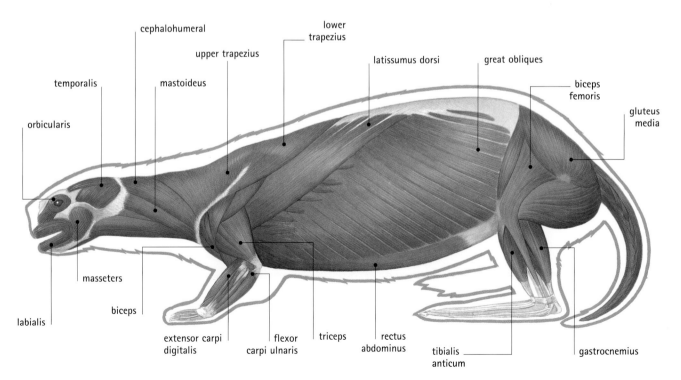

cephalohumeral
lower trapezius
upper trapezius
latissumus dorsi
great obliques
temporalis
mastoideus
biceps femoris
gluteus media
orbicularis
masseters
labialis
biceps
extensor carpi digitalis
flexor carpi ulnaris
triceps
rectus abdominus
tibialis anticum
gastrocnemius

Nervous system

CONNECTIONS

COMPARE the otter eye, which has a lens that changes shape underwater, with the eye of the **SEAL,** which has a large and almost spherical lens that can also focus underwater.

The proportion of brain to body size in otters is similar to that in other mustelids, but slightly smaller than in most other carnivores, such as dogs and cats. The structure and function of the brain in different otter species are closely related to foraging habits, in particular whether the otters feed mainly on fish or invertebrates. In mammals, the part of the forebrain called the prefrontal gyrus contains areas that receive and convey information to specific parts of the body. In otters that are primarily fish-feeders—the river otters and the giant otter—the regions at the front of the prefrontal gyrus that convey motor commands to the mouth and facial area are large and complex. These regions gather information from the otter's whiskers and instruct the mouth to bite and seize the prey.

IN FOCUS

Scent

As in other mustelids, smell is an important form of communication in otters. The deposited feces (solid wastes) of otters are called spraints. Adults leave them in recognizable mounds on features such as logs and rocks, to mark out territories. The musky scent of the spraint tells other otters the age and sex of the local otter, and indicates its readiness to mate. Otters can probably recognize other individual otters by this smell alone. Dominant males mark out the largest or most favorable territories by leaving spraints.

▼ Sea otter
The nervous system of the otter is similar to that of other mammals. Within otter species, there are differences in sizes of certain brain regions, depending on whether an otter uses its paws or mouth and whiskers to hunt prey.

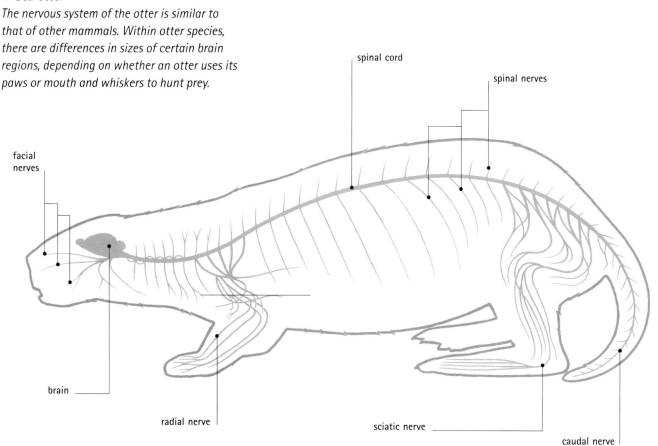

facial nerves

spinal cord

spinal nerves

brain

radial nerve

sciatic nerve

caudal nerve

In the invertebrate-eating otters—the sea otter, African clawless otters, and Asian small-clawed otter—in contrast, it is the front paws which provide sensory information and which gather prey by touch in dark or cloudy water. In these species, it is the regions of the prefrontal gyrus that process and relay information to and from the forelimbs, which are large and complex.

In an early experiment on feeding behavior in otters, a river otter had its vibrissae removed (they do regrow). In clear water and bright light, the animal's hunting efficiency was not impaired, showing that it was then relying on sight rather than touch for successful hunting. In cloudy water, however, the otter took up to 20 times longer to catch prey than it did when it still had its vibrissae.

The eyes of otters in general, and the sea otter in particular, are able to focus well both in air and in water by an apparently unique mechanism. When humans enter the water, their vision becomes blurred unless they wear goggles or a mask. Human eyes are suited for seeing in air. Underwater, the lens in human eyes is unable to bend light sufficiently to focus it on the retina, the light-sensitive tissue lining the back of the eye. Otters have a mechanism whereby the shape of the lens alters dramatically when the animal dives. A ring of muscle around the pupil of the eye contracts, causing some of the clear fluid in the chamber at the front of the eye to flow out into spaces in the surrounding tissue. The fluid pressure at the front of the eye drops, and in response the lens bulges outward. This change makes the lens much more powerful and able to bend light rays more, allowing the otter to focus clearly underwater. This mechanism gives otter eyes the greatest focusing range of any vertebrate. The eyes of most, perhaps all, otter species have an effective tapetum lucidum. This is a layer at the back of the eye that reflects light rays back out of the eye. As a result, the light rays stimulate light-sensitive cells twice—once on the way in and again on the way out—and thus increase the sensitivity of the eye in poor light. This helps otters hunt actively at night, when there is less danger of being seen by predators that could attack them.

EYE IN AIR

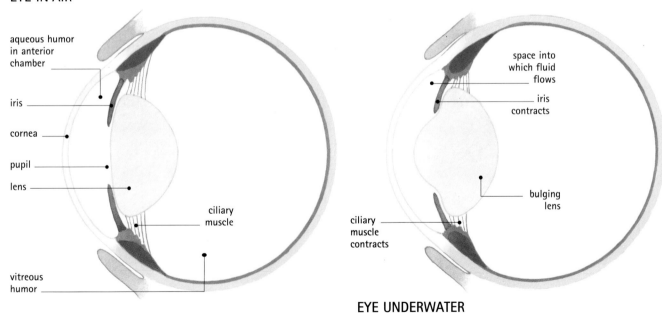

EYE UNDERWATER

▲ CHANGING SHAPE OF THE OTTER EYEBALL
To focus effectively in both air and water—which have different light-bending properties—the lens of the otter's eye can change shape to compensate for this difference. When an otter dives, the iris and ciliary muscles contract, forcing some of the aqueous humor out into surrounding tissues. This causes a drop in pressure in the front of the eye, and in response the lens bulges forward and is able to bend light sufficiently to focus clearly.

Circulatory and respiratory systems

Like all other mammals, otters inhale air through nasal passages, warming it before it travels down the windpipe to the lungs. Sea otters have a particularly complex arrangement of bony extensions called turbinal bones, which block the free flow of air, causing it to swirl around in the nasal chamber, where it is warmed, moistened, and cleaned before reaching the lungs. In the air sacs of the lungs, or alveoli, the fresh supplies of oxygen from the inhaled air are exchanged for the waste gas carbon dioxide. The otter exhales the used air when it surfaces after a dive.

Underwater

Contrary to popular belief, river otters remain submerged only for short periods—typically 10 to 45 seconds for each dive. These otters stay underwater longer only in exceptional circumstances—for example, when they are hunted by predators, such as larger carnivores. Sea otters dive for longer times—typically 50 to 90 seconds—but even these dives fall well short of the dives made by marine mammals such as seals, sea cows, and whales, which may last many minutes or even hours. Sea otters,

being deeper divers than freshwater otters, have a correspondingly wider windpipe, which allows them to breathe air into and out of their lungs more efficiently before a dive. The air passages leading toward the alveoli, including smaller ones, have cartilage in their walls, preventing them from collapsing under high water pressure during a dive.

Keeping warm

Because water conducts away heat much faster than air, otters run the risk of losing too much body heat when swimming. If their core body temperature drops too low, they can die of cold. To combat this, otters are well insulated with fur, which reduces heat loss, and they have a high metabolic rate to generate internal heat.

As in many other mammals, the blood vessels in an otter's limbs are arranged so that arteries carrying blood to the extremities lie close to veins that carry blood the other way. This protective feature, called a countercurrent system, ensures that blood that has been cooled as it passed though the limbs is rewarmed with heat from blood in arteries coming from the animal's warm core.

▼ **Sea otter**
Arteries are shown in red and veins in purple. Otters have unusually large lungs, enabling them to store oxygen when diving.

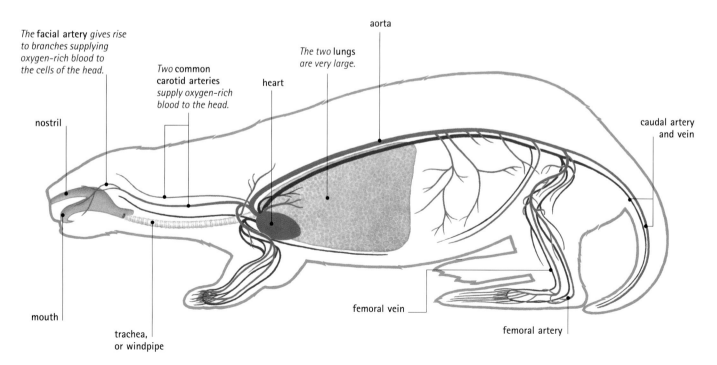

The **facial artery** *gives rise to branches supplying oxygen-rich blood to the cells of the head.*

Two **common carotid arteries** *supply oxygen-rich blood to the head.*

aorta

The two **lungs** *are very large.*

heart

nostril

caudal artery and vein

mouth

trachea, or windpipe

femoral vein

femoral artery

Cooling down

There are times when otters are in danger of overheating. This can happen when an otter has been actively hunting and returns to the water surface, where the air temperature may be very high—perhaps as high as 82°F (28°C) in summer. The otter's thick fur coat increases this risk, too.

The paws of sea otters have a rich network of blood vessels just beneath the skin, and blood is diverted to this network when the core body temperature is in danger of rising too high. The diversion of blood into these blood vessels increases the rate of heat loss from the otter's body into the surrounding seawater.

This same system of blood vessels can work the other way—absorbing heat from the surroundings rather than radiating it. On sunny days, sea otters may sunbathe at the surface with their hind feet extended out of the water, turned to the sun. The feet absorb heat from the sun's rays, warming the blood after a chilly dive in cold water.

▼ *A sea otter sunbathes afloat, anchored to strands of the seaweed kelp to prevent drifting.*

The sea otter's blood contains a concentration of hemoglobin—the blood's oxygen-carrying pigment—that is only slightly higher than that found in land-living mammals of a similar size. However, the otter's hemoglobin is much more efficient at trapping oxygen. This efficient type of hemoglobin enables the otter to store oxygen for use during the dive, when it cannot breathe air. Sea otters also have unusually large lungs—larger, in proportion to body size, than those of any other mammal. This feature enables the otter to store more air for dives. The large air-filled lungs also create plenty of buoyancy, which helps keep the otter float at the water's surface and also counteracts the weight of prey items and stones for breaking shells that the otter collects from the seabed.

The concentration of myoglobin, which is an oxygen-carrying molecule like hemoglobin but found in muscles, is about three times higher in sea otters than in land-living mammals. The otter's myoglobin releases oxygen gradually during dives, allowing the animal to respire and keep its bodily activities functioning when holding its breath for some time.

Like other mammals, otters have a four-chamber heart that pumps blood through a double circulation. In the systemic circulation, arteries with thick, muscular walls carry blood under high pressure away from the heart to supply other organs. Thin-walled veins carry blood back to the heart under low pressure. In the pulmonary circulation, blood takes up oxygen and releases carbon dioxide in the lungs.

Digestive and excretory systems

Most otter species eat a variety of animals, the precise diet usually depending on which prey species are most abundant in the locality. Eurasian and North American river otters prefer slow-swimming fish and fish that seek to escape by hiding rather than fleeing. But they will also take shellfish such as crayfish and mussels and even amphibians, water birds, and small mammals such as baby mice or voles.

Four otter species—the two African clawless otters, the Asian short-clawed otter, and the sea otter—catch mostly invertebrates. They feel for their prey using their sensitive front paws.

The sea otter has an unusual method for collecting and handling invertebrate prey. The otter dislodges invertebrates from the seabed and gathers them in a fold of skin under each armpit that serves as a temporary food store until the otter rises to the surface to eat its prizes. The otter also uses the pouch for carrying a stone from the seabed. On the sea surface, it uses the stone as an anvil balanced on its chest as it swims on its back. Holding larger, hard-shell invertebrates in the front paws, it smashes them against the stone anvil, breaking the shell so it can get at the soft contents.

IN FOCUS

Gaining water and losing salt

Like other carnivores, otters have moderately large kidneys, which they require to flush urea—a waste product from their high-protein diet—out in their urine. Because sea otters need to drink seawater to gain sufficient water for this flushing effect, their kidneys are about twice the size of those of freshwater otters. Seawater has a high concentration of salts, which the sea otter's kidneys must concentrate and excrete in the urine.

Short esophagus

The otter's digestive system is similar to that of other carnivores, with a short esophagus leading to the stomach, where food digestion begins. Juice from the pancreas and bile from the liver empty into the first part of the small intestine just beyond the stomach, continuing the digestive process. Digestive enzymes are also secreted from the wall of the small intestine. The small intestine—so named because of its narrow width—is more than 10 feet (3 m) long and has a folded surface, giving

▼ **Sea otter**
Otters eat up to one-quarter of their body weight each day and have very long intestines and a large liver to process food. They also have large kidneys to cope with their protein-rich diet.

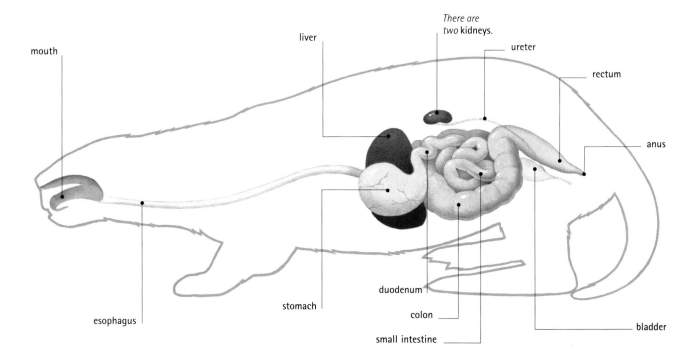

mouth

liver

There are *two* **kidneys**.

ureter

rectum

anus

esophagus

stomach

duodenum

colon

small intestine

bladder

▲ *Eurasian otters use their mouth to seize fish, but once the prey is caught the otters manipulate it with their paws.*

PREDATOR AND PREY

Sea otters, sea urchins, and kelp

In some localities, sea otters play a role in maintaining the health of dense beds of large brown seaweed called kelp. Sea urchins graze the kelp, and in the absence of sea otters, these invertebrates undergo dramatic "boom-and-bust" population cycles. When kelp is plentiful, sea urchin populations grow rapidly until they become so abundant that they exhaust their food supply. Mass starvation of sea urchins ensues and their numbers crash. The temporary loss of kelp in turn affects populations of other invertebrates that depend upon the kelp for food and shelter. When sea urchin numbers are low, the kelp beds grow lush again, and other invertebrates gradually reestablish themselves, until the next urchin explosion. Where sea otters are present, however, the sea-urchin population is kept in check. This helps maintain the kelp beds in a more or less stable state, which is ecologically much healthier.

a large surface area across which products of digestion are absorbed. The liver controls the processing of digested foods absorbed from the gut. Further uptake of digested food and water occurs in the large intestine, with solid waste accumulating in the rectum before expulsion through the anus.

A large diet

Otters eat between 15 and 25 percent of their body weight every day to fuel their high metabolic rate. To process such large amounts of food, otters have an unusually long gut for carnivores and an exceptionally large liver. The gut is up to 10 times the animal's body length, and the liver is 5 or 6 percent of body weight. Most species of otters forage for three to five hours each day to satisfy their high energy requirement. The sea otter, which spends almost its entire life in cold water, typically spends more than six hours a day gathering and eating food.

Reproductive system

CONNECTIONS

COMPARE the delayed implantation of embryos in sea otters and North American river otters with the suspended animation of embryos in the red *KANGAROO*. Both strategies allow young to be born when conditions are favorable.

▼ *Female sea otters usually have just one cub, which they carry on their chest while swimming on their back. The cub develops fast, learning to swim within four to five weeks.*

When a female sea otter is ready to breed, she gives off a scent that attracts males. When a male detects the scent, he swims with his face down among the females, following the chemical trail until he finds the scenting female. He then embraces her while sniffing and rubbing at her body. During mating, which lasts between 15 and 30 minutes, the male takes hold of the female's nose with his teeth. When he finally detaches, the female's nose is red and bloody.

When a female Eurasian otter is ready to breed, she attracts mates by scent-marking the spraint (otter feces) heaps she deposits in her home range. The scent may attract several males, in which case they must fight with each other for the opportunity to mate with her. The winner courts the female, the two swimming close together, twisting and turning around each other. This courtship may last several days before the pair mate. In the Eurasian otter, mating lasts 10 to 30 minutes and is repeated several times over a few days. This triggers egg release and helps increase the chances of successful fertilization.

Gestation

Sea otters and North American river otters have unusually long gestation periods, because of a process called delayed implantation. In these species, the fertilized egg divides into a

ball of cells, but then development halts for several months. Then, the embryo implants in the wall of the womb (uterus) and development resumes. The delay ensures that the young are born in spring, when the climate is milder and more food is available, rather than in midwinter, when the young's chances of survival would be poor. Female sea otters give birth not in dens but usually on the sea surface, often wrapping their body in kelp fronds to stop themselves from drifting. They normally produce just one cub, which they carry on their chest while swimming on their back. Because sea otter cubs are exposed to the elements from birth, they develop much faster than cubs of other species of otters. They are born with eyes open and can swim within four to five weeks.

Eurasian otters commonly shelter in a bankside burrow called a holt for part of each day. Pregnant females carefully select a holt as a breeding den. They choose a burrow close to a rich food source and line the den chamber with leaves, hair, and twigs. The nesting material helps keep the cubs dry while allowing air to circulate through the den.

The Eurasian otter's gestation period is about nine weeks. The mother gives birth to between one and four cubs, which are born blind, toothless, and almost helpless. In the first few weeks, their development is quite slow, even though the mother's milk has a high fat content

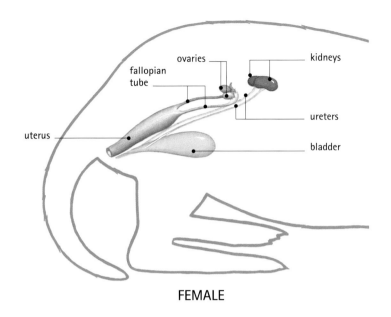

FEMALE

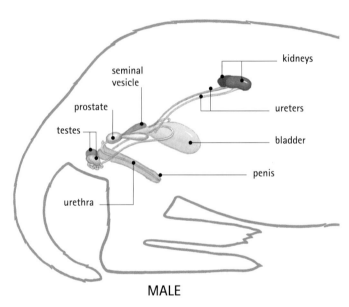

MALE

Sociable or solitary

The sociability of otters varies depending on species. Eurasian otters are largely solitary as adults, coming together in pairs only during the breeding season. Larger groups tend to consist of only a mother and her cubs. African clawless otters and North American river otters, on the other hand, are more sociable. Outside the breeding season, males commonly live in loose groups of a dozen or more otters. In addition, female sea otters live in large groups.

Male Eurasian otters typically play little or no part in rearing the young, although they may occasionally accompany the family for a few days. The mother may take more than a year rearing her cubs so they are able to catch fish and fend for themselves. Training her young to hunt fish involves releasing prey she has caught so that her young can recapture them. Young otters chase and play at fighting for much of the time, preparing themselves for adult life when they have to actively hunt for food and, as males, may have to fight for a mate. At about one year old, young otters leave their mother's territory and seek to establish a home range elsewhere. Otters become sexually mature by about two years old, but many succumb to starvation or are caught by larger carnivores and birds of prey before they reach maturity. In the wild, Eurasian otters rarely live to eight years old.

TREVOR DAY

FURTHER READING AND RESEARCH

Kruuk, Hans. 1995. *Wild Otters: Predation and Populations*. Oxford University Press: Oxford, UK.

MacDonald, David. 2006. *The Encyclopedia of Mammals*. Facts On File: New York.

Perrin, W. F., B. Würsig, and J. G. M. Thewissen (eds.). 2002. *Encyclopedia of Marine Mammals*. Academic Press: San Diego, CA.

Reynolds, John E. III, and S. A. Rommel (eds.). 1999. *Biology of Marine Mammals*. Smithsonian Institution Press: Washington, DC.

▲ REPRODUCTIVE SYSTEMS

Sea otter

The female otter has a long tubular uterus (womb). Female sea otters usually give birth to just one cub. Male otters of all species have a bone in their penis called a baculum (not shown).

(about six times that of cow's milk), and the young suckle every few hours. The young do not begin to take their first solid food until about seven weeks old, which is about the same time that they take their first faltering steps. During this period, the mother has to hunt regularly to sustain her production of milk.

Eurasian otters are not fully weaned until about 14 weeks old, by which time they have developed their first waterproof coat of fur and have learned to swim. Soon after, young Eurasian otters begin catching their own food, beginning with slow-moving invertebrates rather than faster-moving fish.

Index